Alternative Ecological Risk Assessment

"Truth will sprout from earth, and righteousness will peer from heaven."

Psalms (85: 12)

". . . how many times can a man turn his head
Pretending he just doesn't see?"

Bob Dylan ('Blowin' in the Wind')

Alternative Ecological Risk Assessment

An Innovative Approach to Understanding Ecological Assessments for Contaminated Sites

Lawrence V. Tannenbaum

Army Institute of Public Health
U.S. Army Public Health Command

WILEY Blackwell

Registered office: John Wiley & Sons, Ltd, The Atrium, Southern Gate, Chichester, West Sussex, PO19 8SQ, UK

Editorial offices: 9600 Garsington Road, Oxford, OX4 2DQ, UK
The Atrium, Southern Gate, Chichester, West Sussex, PO19 8SQ, UK
111 River Street, Hoboken, NJ 07030-5774, USA

For details of our global editorial offices, for customer services and for information about how to apply for permission to reuse the copyright material in this book please see our website at www.wiley.com/wiley-blackwell.

Library of Congress Cataloging-in-Publication Data

Tannenbaum, Lawrence V.
Alternative ecological risk assessment : an innovative approach to understanding ecological assessments for contaminated sites / Lawrence V. Tannenbaum.
pages cm
Includes bibliographical references and index.
ISBN 978-0-470-67304-1 (cloth)
1. Hazardous waste sites–Risk assessment. 2. Ecological risk assessment. I. Title.
TD1052.T36 2014
363.72′872–dc23

2013022392

A catalogue record for this book is available from the British Library.

Wiley also publishes its books in a variety of electronic formats. Some content that appears in print may not be available in electronic books.

Front and back cover design by Andrew Leitzer.
Cartoon illustration by Joyce Kopatch. Deer photo by Ryan McVay, collection: Digital Vision. Forest Image © iStockphoto. Fox photo by Jupiter Images, collection: liquidlibrary.

Typeset in 10/13pt RotisSemiSans by Laserwords Private Limited, Chennai, India
Printed and bound in Malaysia by Vivar Printing Sdn Bhd

1 2014

To my wife, my *aishes chayil*, Chava Esther –

I could only have met the task and succeeded with your support, constant encouragement, and advice.

May we continue to see our beautiful family grow.

Avi

Contents

Preface

There are two sides to every story (or maybe more than two sides). There are different ways to look at situations and different ways to understand phenomena going on around us. It is not only in the realm of interpersonal relationships to which the adage applies; science and its application are areas that are ripe for seeing and explaining things differently. And so this book ... Its purpose is to present a different side of the story for ecological risk assessment (ERA) as applied to contaminated sites that proceed through a process intended to determine if receptors are in harm's way and to take appropriate action if they are.

Having read just this much, one might suppose that the author was looking for a still different angle of ERA on which to write, and perhaps looking to be contrary for the sake of being contrary. That's not the case, however. It's a sad commentary when one has to resort to developing a contrary opinion or understanding in order to arrive at a topic that no one has yet secured. I would most definitely not encourage others to set themselves down to the specific task of finding flaws or shortcomings for the particular scientific field in which they work, all for the purpose of honing in on a topical area that they can uniquely claim as their own. For the book you are holding (dare I say 'kindle-ing', because along with the reference to the trendier means of reading books today, the suggestion could be that the subject book might best be suited for trashing in the fire), the etiology is much more straightforward and genuine. There was no deliberate process of systematic review of ERA for the express purpose of uncovering faults and the like so as to have a sufficiency of material to fill the pages of a book. The material was instead assembled in a casual manner over the years as the author worked at his science.

To me, the author, the observations and analyses I set forth are incredibly obvious and unmistakable. The staunch supporter or defender of common ERA practice will feel quite differently though, and that's fine, for science grows from healthy constructive debate. Perhaps I am completely wrong about every ERA element the book touches on. A reason to think so is that I appear to be the only one giving voice to our having an ERA process that leaves so very much to be desired. Perhaps we are so ingrained in our way of thinking that we never pause to consider other possibilities, and perhaps I am the only one who did pause to so contemplate things. There is the matter too, of not wanting to hear an alternative understanding of it all, and to complete the spectrum, there is the vastly more serious matter of not allowing oneself to hear an alternative analysis, a topic certainly dealt with in the book.

This book's purpose of presenting an alternative understanding of ERA is set forth with the hope that it will pique the minds of those who, in one fashion or another, are involved with ERA as an outgrowth of ecological science. The book is expressly targeted for the serious (college or graduate) student of health risk assessment, environmental science, or ecology, those who work in the ERA field, and those who publish on ERA topics. It should be required reading for regulators who craft and dispense ERA guidance and policy, and who need to hear among other things, why an ERA process isn't needed, not that we have one anyway – alas, not an avant-garde concept to summarily dismiss, but merely an alternative concept to professionally consider.

Lawrence V. Tannenbaum
September, 2012

Acknowledgments

I offer my great thanks to the US Army Institute of Public Health (AIPH; formerly the US Army Center for Health Promotion and Preventive Medicine, and before that still, the US Army Environmental Hygiene Agency) for allowing me, where they could, to test the waters – to conduct studies critical to a bettered understanding of the potential for chemicals in the environment to impact ecological receptors. Special thanks to Dennis Druck for guiding my career especially through turbulent times, and for nurturing my passion to unleash a different perspective so clearly borne out by the studies. I am indebted to the US Army Environmental Command (formerly the US Army Environmental Center) and to Jim Daniels in particular, for funding one-of-a-kind studies that bear up to scrutiny and that are secured in the peer-reviewed literature. Several Army installations provided the necessary funds to carry out work I so badly wanted to have done. I send my appreciation therefore to Picatinny Arsenal (and Ted Gabel), Badger Army Ammunition Plant (and Joan Kenney), and Joint Base Langley Eustis (formerly Fort Eustis; and Joanna Bateman). The list of individuals to thank for allowing RSA (see Chapter 9) to be birthed, cultured, and firmly established, is extensive. There were thinkers, those who supplied the grunt work, and more. The short list runs to: Keith Williams, Brandolyn Thran, Adam Deck, Jeff Leach, Sue Fox, Barrett Borry, John Buck, and the techs from AIPH's Soils Lab.

1 An introduction and overview

At the time of writing, I have over 20 years of experience working in the health risk assessment field as it relates to chemically contaminated properties, with a special emphasis placed on ecological risk assessment (ERA). At the beginning of my career in environmental work while with the United States Environmental Protection Agency (USEPA), I was involved with the pre-remedial element of the legendary Superfund program. My duties there largely amounted to verifying the conclusions recorded in preliminary assessments (PA) and site inspections (SI) for the contaminated sites. Most of the documents recommended that the sites advance to more refined tiers of analysis that would permit non-threatening sites to be shifted away from those that seemingly had the potential to truly pose substantial health threats to the human and ecological receptors that contacted them. I'm not certain but it may be that already back then, I pondered if we were ever encountering instances of harm in the ecological receptors that inhabited the sites which were producing substantial scores with the Superfund program's Hazard Ranking System (HRS). In my reviews I had never come across a single PA or SI that had reported tell-tale signs of injury to birds or mammals, the two groups of terrestrial ecological receptors that are routinely evaluated in ERA work. None of the PAs or SIs had described the unsettling discovery of a site devoid of biota, and at no site or anywhere nearby to one, had anyone observed something bizarre and clearly out of the ordinary, such as an accumulation of carcasses or evident signs of rampant disease in an ecological (i.e., non-human) species. To a certain extent, I shelved for a while the need to have my query resolved. If there really were health concerns for site ecological receptors, unaware of these as any of us might be, I reasoned that such would be indirectly captured and later addressed courtesy of the HRS scores. Sites that would advance through the process (and perhaps all the way to the National Priorities List; NPL) due to their high scores, would be there via a ranking scheme that was heavily weighted to human health concerns. Once on the NPL though, the sites would come under greater scrutiny not only for their potential human health concerns, but for any ecological concerns they might bear as well.

Alternative Ecological Risk Assessment: An Innovative Approach to Understanding Ecological Assessments for Contaminated Sites, First Edition. Lawrence V. Tannenbaum.

About 10 years into my risk assessment craft, having already logged in a number of years with the Army pursuant to my EPA years, and having moved well beyond attending to pre-remedial tasks only, the nagging thoughts returned. For all that I knew of contaminated terrestrial and aquatic sites managed under the related Superfund and Resource Conservation and Recovery Act programs (i.e., valuable and relevant knowledge gleaned from hands-on involvement with the sites, and not just information assimilated through reading work plans and remedial investigations and the like), I could not point to single site that had anything ecologically wrong with it (e.g., vanished populations or species or a lacking detritivore compartment). Site visits didn't suggest anything amiss, and through negotiations with stakeholders of all walks, many insistent that their sites submit to extensive eco-based site cleanups, never once was it suggested that there had been observations of ecological health compromise occurring in the field. [A clarification is in order here. Site visits can very well detect (gross) habitat loss or elimination, as in the "slickens" of toxic waste sediments (i.e., the highly compromised riparian areas) of the Upper Clark Fork River Superfund Site in western Montana, that are directly traceable to the mining, milling and smelting activities that have occurred over many decades (Brumbaugh et al. 1994). Where such massive physical destruction has occurred though, it's understood that the lack of habitat is responsible for the removal of the resident biota of yesteryear. Importantly then, construction projects to restore habitat are the order of the day, and moreover, riparian species aren't at risk. With the habitat they require missing, riparian species are not expected to be present and thus at risk from chemical exposure. It bears mention here too, that while individuals might be quick to cite the Upper Clark Fork River Superfund Site as one having been ecologically wronged, the example is a poor one. This site, the largest geographic Superfund site in the United States, does not typify in the least the sites that are the subject of this book. The overwhelming majority of sites, Superfund or otherwise, are just a handful of acres in size, if that much.

In all fairness, there was no need for the above-mentioned stakeholders to have arrived at such a point about (lacking) observed impacts. For Superfund sites and for others that defaulted to Superfund guidance in the absence of having their own program-specific guidance to follow, it was always a risk-based framework that was in place. Aptly then, the focus was always on what *could* happen to the ecology should a site remain as it was, i.e., in a non-remediated state. Program dictates aside though, I realized that I had nevertheless happened upon a key distinction between the human health risk assessment (HHRA) and ERA processes (discussed below). This distinction in focus, largely centered on spatio-temporal considerations, was eluding the rank and file of health risk assessors. Perhaps it was because I was a risk assessor for both human health and ecological health when most people in the health risk assessment field did just one or the other, that I was able to secure and frame my observation. Perhaps too there were other reasons why the ERA practitioners were not able to do the same. Surely they'd had ample opportunity to become acquainted with the distinction, for in more recent years I had dosed the ERA landscape with my notion via published peer-reviewed pieces and speaking engagements at professional society conferences. My sense is that ecological risk assessors and ecotoxicologists have gone out of their way to turn a deaf ear to the matter. Although I haven't yet taken the time to prove it, it may be that the very notion – in its ultimate manifestation, that ERAs are unnecessary – repulses this audience; hearing that notion articulated is, at the very least, perceived as a threat to job security. Let us explore the notion, which to my way of thinking is as glaringly obvious as the one essential difference between HHRA and ERA that ecological risk assessors *do* acknowledge, namely that HHRA evaluates just one species, whereas ERA is left to potentially consider all other species that inhabit a site of interest.

On any given day, the prospect of there being a human newcomer to a contaminated property can be realized. Human beings by nature relocate, and frequently do so over considerable distances. This occurs for example, when a family breadwinner accepts a job transfer, triggering a family move out-of-state. Such relocation or repositioning creates new opportunities to arrive at and consistently occupy (through residence or employment) what may later be found to be a contaminated address. It is for this reason that by convention, HHRAs use assumed exposure durations of 25 years for industrial and other workers, and 30 years for residents, i.e., 90th percentile figures for site occupancy that reflect considerable research into human activity patterns (US EPA 1989). So real is the prospect of humans arriving anew to a residence or workplace and to have their full *run of it* there (i.e., to realize such 90th percentile exposure duration terms), that consideration is never given to those who have already been living or working at the site for some time.[1] The salient point is that at best, only a portion of the lifetime of an individual or a population is ever considered for health effects that might crop up; there is no concept of assessing the progeny of those unlucky individuals who might have arrived anew at some contaminated address, and with good reason. Although there may be faint vestiges of the extended family in our country today, where multiple generations live under one roof, for all intents and purposes we do not encounter such site settings. As mentioned earlier, humans

[1] To my mind it is a most unfortunate happenstance that conventional HHRAs for hazardous waste sites do not entertain such exposure scenarios. What though, of the worker employed for the past several years at an industrial park where soil ingestion exposures or inhalation exposures of the site's wind-entrained contaminants have only today been discovered to be unhealthful? We should surely want to know the risks for those seasoned workers who fortunately have not yet presented with illness, and especially when this population is seemingly closer to presenting with illness than are newcomers to the site arriving as you read this! Exploration into these and other neglected scenarios could lead to a utilitarian expansion of the science of HHRA as we currently have it, enhancing our understanding of thresholds-for-effect, latency of effect, and developmental disease.

relocate and reposition. Grown children might find themselves living in the same general area in which they matured, but it's more than highly unlikely that they would be living at the very same contaminated addresses they always knew. Risk assessment's guiding principle of reasonableness (in considering exposure scenarios to evaluate, etc.) would not be met if we were to *de facto* include in our HHRAs, the children, grandchildren, and great grandchildren of the "new arrival" residents and employees described here. What about ecological receptors? Firstly, the ones that might concern us at Superfund-type sites simply do not arrive at settings anew, to experience exposure durations of only singular lifetimes. Additionally and in stark contrast to humans, they do not undergo geographical repositioning on a scale of such magnitude, or at least not in a manner that would exacerbate their degree of contaminant exposure for any given site of interest. A clarification with regard to repositioning over vast reaches is in order here, for many birds are of course, migratory in nature, overwintering at locations that are hundreds and perhaps thousands of miles distant from the areas they occupy in the summer months. The more time a migratory bird is away from a site however, the less appropriate it is for inclusion within an ERA. It is also worth noting at this juncture that, for what should be some rather obvious reasons (discussed later), a bird that does not reside at a contaminated site for a stretch of three or four months or longer is a rather non-desirable receptor of concern to evaluate altogether.

Reduced geographical repositioning in ecological receptors is best demonstrated with an example from mammals. A raccoon or a local population of raccoons will not appear tens of miles away from where once sighted. Raccoons, like all animals, are bound by their biologically dictated habits and behaviors. In a worst-case scenario, this might very well mean that not only raccoons, but a great many other ecological species are effectively *trapped* at contaminated sites of interest, not having the ability to get beyond the site's boundary. In that particular arrangement where a contaminated site is sufficiently large relative to a receptor's home range, the receptor's entire life is lived on contaminated turf. Now we must ask how long do ecological receptors live? For our considered raccoon, a maximum figure might be just over 3 years, with sexual maturity kicking in at shortly past 1 year. And now we must ask how long have sites been contaminated by the time that they come up for assessment within Superfund or any other similarly structured program? The answer here is multiple decades, with 30 years as a vetted figure. Combining this information we find that any raccoons observed at a contaminated site today are, with only minor exception, descendants that are some 10 generations or more removed from the raccoons that inhabited the site when it first became contaminated.

The implications of the above are far-reaching and they set us on an entirely different plane from that on which we routinely express concerns about human exposures to contaminated sites. Whereas the prototypical modeled hypothetical humans of HHRAs are consistently assumed to have had no prior site-specific contaminant exposure, a site's ecological receptors have only known life within a contaminated context.[2] For those who sense the urge to immediately challenge this because immigration is an undeniable element of a dynamic ecosystem, let them consider two other mainstay species of ERAs that could never have known life elsewhere: earthworms and field rodents. When we encounter these species, well-known to be severely restricted in their horizontal migrations, we do not suppose that they are site newcomers, having recently been air-lifted to the site in question through some revolutionary plot to re-introduce species to areas from which they may have vanished for one reason or another. And so, if we observe any raccoons, earthworms, or field rodents at sites today, some 30 or more years after contaminants have been released, don't we then "see" (effectively "know") that the site is still supporting these species? Several other questions need to be asked now: Could there still be a need to assess raccoons at the site? Isn't the passing of 30 years and 10 generations more than enough for a contaminated site to have elicited an observable toxicological response in the ecological receptors that are "trapped" there? Is anyone suggesting that we wait still longer to see if site ecological receptors ultimately present with observable signs of disease or until their populations plummet?

It might be critical here to acknowledge that ERA practitioners are not known for bemoaning the fact that nearly all the studies that support Toxicity Reference Values (TRVs) are one-generational in nature (Sample et al. 1996). Whether they realize it or not, the masses working ERA are not troubled with developing TRVs from such cohort studies that rarely exceed 1 year in duration. They illustrate this posture when they summarily and confidently conclude that ecological receptors bearing hazard quotients (HQs) greater than 1.0 are at risk for developing deleterious health effects. If *we* (the author quite obviously, not included in this collective "we") can find merit in employing such TRVs that reflect the toxicological responses of only previously non-exposed animals to a chemical dosing regimen administered over a portion of their lifetime, we should be nothing less than ecstatic to have at our disposal, site fauna that reflect decades and multiple generations of exposure. If we are willing to see it, we have the opportunity at both terrestrial and aquatic sites, to evaluate the very receptors that have been chemically exposed for more generations than any laboratory study could ever rival. An alternative understanding or appreciation of the contaminated site dynamic is in order then. Currently for a site that presents with a suite of contaminants in its media, such as a field where tanker-hauler trucks have had their

[2] I am reminded of a naysayer, who remarked after a conference presentation I had given, that at a certain contaminated site, a new rookery had only recently been established. I can accept this and other similar occurrences, but I think it only fair to point out how uncommon it is that a site is sufficiently large to support a rookery, a point to be discussed later.

rinsate poured onto the ground over many years, we see it as our job to decipher the potential health effects that the site stands to pose to the unsuspecting biota situated there. A reframed perspective recognizes that when employees of the operation began repeatedly and carelessly allowing the rinsate applications to occur, unbeknownst to them, a toxicity study of exceptionally high quality was getting underway. Unknowingly, the well-meaning employees were initiating a long-term dosing study to the indigenous biota, and often with a well-defined chemical load. In this "study", no animals had to be procured from suppliers nor did they need to be randomized to cages or tanks. No feeding duties had to be assumed, and temperature and lighting settings were a non-consideration. Nature ran the study, and importantly we have the luxury today of reviewing what's happened in the aftermath of a dosing regimen conducted in a natural setting and for a duration that puts any lab-based experimental study design to shame. Consequently, it is far too late to be asking about what *could* happen to the ecological receptors that populate the site. *Risk* assessment for the many sites that share similar histories to this one is no longer what's needed.

More than a decade ago I had yet a third brush with the notion that ecological receptors within a Superfund-type site context fail to bear tell-tale signs of harm due to their chemical exposures. This third occasion profoundly influenced my career path, spurring me on to allocate my available time to exploring the phenomenon of absent health effects in ecological species, with an emphasis placed on elucidating the basis for the absenteeism. In a large way, this book is the product of those explorations that bear on both theory and empirical science. On this fateful third occasion, I was arranging a slide presentation for an ERA course module that I would be teaching. For the first slide of the module I wanted to have a montage of images of the health effects with which ecological receptors contacting or residing at contaminated sites present. I had trouble composing the slide, and could do little more than pull in an image of a cross-bill condition in a gull from the Great Lakes region, and a second image of a frog with a clear case of polymelia (extra limbs). Whereas I wanted a slide that would effectively ring out with "These are the sort of things you can expect to see happen at contaminated sites, and this is why ERA is so necessary", I found that I had hit a brick wall. I thought back to the hundreds of ERAs and related documents I had reviewed over the years. Never had there been a description of what chemically wronged ecological receptors looked like. In my sudden, and fortunately only momentary naivety, I reasoned that ERA boiled down to ecological receptors simply dying off at contaminated sites, this occurring before somatic changes in a critter's external appearance would arise. This would explain why I was having such difficulty in assembling examples of physically altered or deformed organisms. Perhaps then, all I needed for that elusive first slide were two pictures taken from the same contaminated site, arranged side-by-side, one with a bunch of birds or fish, and one with just a few specimens of these forms (the latter presumably indicating the damage caused by the site over a period of years). But if concern over population loss was what ERA amounted to, where were the accounts of dwindled down or decimated populations at the sites? Surely time enough had elapsed at sites for it to be evident that decidedly lesser numbers now populate them relative to either former times or to what the textbooks have to provide in the way of species density figures. Returning to my senses, I questioned if the prototypical sites being addressed by Superfund actions were even large enough to house sufficiently large populations of any species worth assessing. As the reader will recall, prototypical sites still to this day are of the 5- to 10-acre genre with many far smaller than this.

Still endeavoring to compose the slide, I researched the cross-billed gull photo only to learn from the US Fish and Wildlife Service biologists who had taken it, that the condition was anomalous, if not specious. For approximately a decade, the condition had been observed in the Great Lakes region. It had recently disappeared, never allowing for a good understanding of why the phenomenon had come about altogether. In truth, such a condition even if directly related to contaminant exposure, would *not* have been germane to ERA. Superfund-type ERA sites, the interest of this book, are not the size of US states or of still larger regions encompassing multiple states. Further, no formalized process exists for assessing ecological risk for receptors over such a grandiose scale as the Great Lakes region. Real as the time-limited cross-bill condition in the Great Lakes had been (Giesy et al. 1994), and real as other cross-bill instances that have crept up over the years may have been, never has it been suggested that a singular "site", such as a 10-acre lake, was the causative factor. True as it may be that such compromised birds ultimately succumb from undernourishment due to an impeded ability to consume their foodstuffs, such situations call for forensic ecology and not ERA. Once a toxicological response has been elicited, and certainly when one takes the form of a clearly observable physical manifestation, risk, the probability of a manifestation *arising*, is not what needs to be assessed.

For the polymelic frog image I had intended to use for the slide, I encountered issues similar to those with the cross-billed gulls. News briefs and peer-reviewed articles at the time reported discoveries of classical frog deformities and malformations at an alarming clip. Additionally, causation had already been explained and handily demonstrated in multiple ways, with a shrinking atmospheric ozone layer, and limb bud invasion by parasitic cysts serving as the two triggers dominating the peer-reviewed literature and the growing scientific discussion of the day (Cohen 2001). A flood of northeastern US states reported the phenomenon, with Minnesota garnering perhaps the greatest degree of attention. In all of this, no one was suggesting that environmental contamination was the root cause of the problem, not when these effects appeared over such an extensive area. To suggest then, that a specific water body on a Superfund-type site had the potential to promote polymelia, polydactyly, or any of the related deformity types, made even less sense. Not surprisingly I found it curious several years ago, when colleagues of mine working on ERA issues for a Minnesota lake, had brought lake water samples into the lab so that the FETAX assay (Dumont et al. 1983; ASTM 1991) could be run. With the site located in Minnesota, a region we might term the "the anuran deformity capital of the country", what on earth were my colleagues after? The weaknesses of FETAX put aside for the moment (Tannenbaum 2008), were they possibly thinking that a negative assay outcome for the lake water would be an indication of eco-unacceptable contaminant levels? How could one *not* expect to have a failing FETAX assay for waters taken from just about anywhere within the "Land of 10,000 Lakes"?

As for that troublesome first course slide, although neither the frog nor the cross-billed gull reliably depicted what resulted from contaminant exposures, I decided nevertheless to retain the images and to instead modify my talking points. Those images along with an explanation of the near impossibility of locating any others illustrating health-challenged ecological species at contaminated sites, served me well for a different course module, one on the questionable need for an ERA process altogether.

In the foregoing, I have alluded to the distinct possibility that ecological receptors at conventional Superfund-type sites are incapable of presenting with signs of stress or impact, and I have provided just a smattering of reasons why this may be so. I have also suggested that risk assessment is not that process that we need to employ. Understandably these are not trivial suggestions, and more than likely, you the reader, will hear them as nothing less than pointed allegations that the field in which you work can be challenged on many scores. Nevertheless, if we profess ourselves to be honorable stewards of science (and I trust we do), drawing upon a wealth of disciplines to include (as a short list) biology, ecology, ecotoxicology, environmental chemistry, and of course, risk assessment, we need to hold ourselves above all bias. We need to be willing to entertain notions that we haven't to this point, even if they should run counter to what we may want to hear.

Identifying instances of harmed receptors at sites that concern us is a good place to start if we are to make inroads into entertaining alternative understandings of the contaminated site dynamic. Realistically, the more difficult it is for any of us to identify instances of demonstrated harm, the more credibility there is to what I suggest. Recently a professional colleague challenged me when I spoke about the absence of harmed ecological receptors at contaminated sites. I asked him for examples of harm or damage occurring to site biota that he knew of, and the list he rattled off left me feeling ever more confident that there is purpose in my coming forward with this book, a re-examination of various elements of what we refer to as ERA. All five sites on his list were aquatic ones, and huge ones at that.

I found this curious given that not long before our exchange, this individual had published a paper discussing the non-feasibility of conducting ERAs for enormous, watershed sites. Leaving that point aside, the first three sites listed were mining sites, for which the challenger prominently noted that it remains unclear whether it is toxicity (from mining wastes) or habitat disturbance that accounts for the lesser fish populations that have been consistently noted. The remaining two sites on his list were river sites, and also made for a let-down for me; a "let-down" in the sense that my colleague believed whole-heartedly that he had furnished me with clear-cut proof when, in actuality, he had not. How could he have thought that he had furnished me with proof, I rhetorically ask, when for the first of these two sites, he was merely relating the conclusions drawn by "the U.S. EPA and other groups"? I wanted to hear of this individual's first-hand knowledge of death, doom, and destruction, and instead he was telling me what others felt. If that wasn't bad enough, his information was rendered completely worthless when those second-hand conclusions were considered for just a moment. The information my colleague had related was that benthic invertebrates and shrews were being affected by polychlorinated biphenyls (PCBs) at the sites. It seems my colleague overlooked the reality that in our field, cleanups are not pursued for the purpose of affording protection to benthic invertebrates or shrews. And while on the subject of those species, I didn't get the sense that anyone from "the EPA or the other groups" had seen health-compromise in these species with their very own eyes. Regulators rarely venture out to contaminated sites, and I've never known a regulator or any other interested party to scout out a site in the middle of the night, something that would have had to be done in order to view any nocturnal shrews. It was rather clear that the conclusions my colleague had related were based upon computed desktop hazard quotients (HQs), something most unfortunate and something I should have expected. (Among numerous shortcomings discussed later in the book, HQs do not express risk and they routinely assume values greater than 1.0 for even the most pristine of settings.)

Regarding the previously mentioned let-down I describe, I freely admit that I was at the same time, quite fine to see that a mover and shaker in the ERA field had not been successful in deposing me when given the chance. Perhaps then I am quite correct in my thinking. That said, I am not so naive to believe that there exist no cases of health-compromised biota in contaminated site settings. In fact, I freely admit to knowing of a few (or more correctly, a very few), such as selenium-posed kyphosis and lordosis (spinal curvature aberrations) in fish exposed to wastewater released from a coal-fired electric generating facility (Lemly 1999). As far as terrestrial settings go, I still don't know of any examples of health compromise, and I wonder (doubt?) if you do. The reader is reminded that aside from a "failing" HQ not constituting proof of stress or of compromised health, neither are elevated tissue concentrations in organisms. Mere chemical presence in living tissue or in any other environmental medium is not risk and does not illustrate harm. For there to be risk, chemical presence must be related in some meaningful way to the onset or development of health effects.

More than simply alluding to topics that this book will address in later chapters, an essential point I have striven for in this introduction is that it is not only the question of "why conduct an ERA altogether?", that is suited for re-examination. If we are to apply the best science, we need to be willing to consider that our present understanding of all ERA components might be in critical need of modification. It troubles me greatly, but I seriously suspect that ERA practitioners would be highly disinclined to hear that they need to shift gears in order to be more in step (or in step at all) with the science of their field. I am troubled too, by what I think constitutes the greatest of impediments to improving our ERA lot, namely that we are afraid to hear "good news"; that although sites may be contaminated, the ecology is no worse off because of it. If our approach (or perhaps, mantra) to ERA smacks of "a contaminated site, by definition, cannot help but pose health challenges to site biota", we have otherwise set ourselves up for a self-fulfilling prophecy. We will be sure to find a way to convince ourselves each and every time, that indeed there are problems rampant at our sites of interest. I am reminded of a debate over the merits of modeling in ERA, in which I participated not long ago. Not surprisingly in my argument about the failings of modeling, I had rattled off a short list of indisputable HQ method limitations, these from a longer list that I have popularized in the peer-reviewed literature (Tannenbaum et al., 2003b; Tannenbaum 2005a). My opponent rejoined with a statement to the effect that (HQ-based) ERA models have been around for some 25 years, and have therefore served a useful role since there appearance (Barnthouse 2008). How saddening, I thought. Does the fact that a practice has been in place for some period of time grant attestation that the practice is a correct or useful one? I trust that you answered "no" to the question, and recognized it as a rhetorical one. To illustrate this matter of blind allegiance to a tool that we may have at our disposal though, I will just touch here on something that will be discussed later at greater length in the context of a case study.

For 10 years, a model has been in vogue that estimates the likelihood of a bird incidentally ingesting a spent lead shot pellet lying on the ground (Peddicord and LaKind 2000). Does the decade's worth of model availability and use bestow credence and utility to it? I wouldn't think so, and I would hope others would similarly respond. If you answered that the decade's worth of availability does indeed furnish proof of its utility, I would then ask if you'd answer the same after learning that the published model contains a mathematical error (Peddicord 2010). Only recently discovered, we now know that bird grit ingestion model users have been generating erroneous values all along (as in US FWS 2002), and because of that, equally erroneous figures have been generated for pellet densities atop soil that are assumed to be safe for local bird populations. I argue that just as this model has been generating erroneous values without our knowledge and for so long, perhaps too, what so many may have accepted as givens in conventional ERA lore are also tainted. It is entirely possible that ecological receptors are insufficiently exposed to chemicals such that toxicological effects don't ensue – and thus our understanding of exposure assessment might need correcting. It is entirely possible that chemicals as they exist in the field are not as toxic as we think – and thus toxicity assessment, for a great many reasons, is due to be revamped. It is entirely possible that ecological receptors can successfully adapt, genetically and in other ways, as a means of living amidst contamination. The recent discovery that Atlantic tomcod (*Microgadus tomcod*) in the Hudson River, in only 50 year's time, developed a genetic mechanism through which toxicity posed by PCBs is dramatically dampened, is a glaring case in point (Wirgin et al. 2011). Finally, it is entirely possible that we are ERA's greatest enemy, thwarting the development of much improved science because we will not relinquish our stronghold on a process that is embarrassingly simplistic and fails to "assess" ecological risk altogether. In the ensuing chapters, please allow yourselves then to consider an alternative understanding of the ERA process and its various components as the process is applied to contaminated Superfund-type sites.

As the hopefully enthused reader sets to the task of digesting the chapters yet to come, the author finds it appropriate to provide some brief discussion here on what to expect, and some general pointers too, on how to proceed. First and foremost, it is recommended that the book be read completely through, and in sequence. As per the book's subtitle, it was the intent to gradually assemble the "innovative approach to understanding ecological assessments for contaminated sites"; while there is value to be had nevertheless from digesting an isolated chapter, as in an assignment to a student, the fully constructed argument to be solidified may be somewhat jeopardized. The absence of subchapter headings then is deliberate, and does not reflect laziness on the author's part. The writing style is intended to facilitate the flow of a consolidated and holistic discussion. Additionally, because the book does not regularly deal with subtopics or equations and the like, there was much less need to compartmentalize, and thereby, resort to the use of subheadings.

This book is not a rant about the current ERA process, but in order to work through the concepts of an alternative understanding of ecological assessment, acknowledging the unfortunate design of the reigning ERA process is an indispensable first step. It is imperative that we do not allow ourselves to be impressed with what the current paradigm has to offer. Admittedly the book bears numerous references to the HQ and unquestionably to excess, but these are all necessary. The HQ is both the epitome of the current paradigm and the "archenemy" of those who would like to see ecological assessment for contaminated sites advance. The excessive attention the HQ draws as part of the current paradigm is singularly responsible for blocking better science from coming about. At times, the text resorts to the use of hyperbole. While efforts were made to reduce this, what remains is placed there for emphasis.

On numerous occasions you will read about "ERA practitioners", "the powers-that-be", "movers and shakers", "the open-minded or engaged reader" and "the naysayers". Undeniably, there is considerable polarization among those working in the ERA field, and the interaction among parties is not always pretty. To not discuss the science in a context of the perceptions and beliefs of those who approach the various process elements so differently, would be an injustice.

To the extent that the reference lists are somewhat brief, more than anything this reflects the forward thinking that is presented. Published references for the newer concepts discussed are not readily found in the open literature. Even before that though, we do not find a great literature or even a minor one, directed at critiquing the established ERA process. Where some might think that sufficient attention has already been thrown to the enhancements the ERA process could know, as in Dale et al. (2008), or to descriptions of a new paradigm for risk assessment, as in Forbes and Calow (2012), this is simply not the case. Overall, it would be presumptuous for the author to frequently self-cite in order to enlarge the references list. Finally, to the extent that the book finds the current ERA paradigm to be wanting – and to the point that alternative assessment approaches need to be entertained – judicious thought should be applied when assigning book reviewers. It is most difficult to procure objective review from those who find a process they wholeheartedly endorse having been severely critiqued and challenged.

References

ASTM (1991) Standard guide for conducting the frog embryo teratogenesis assay-Xenopus (FETAX). ASTM E1439–91. In: *Annual Book of ASTM Standards.* American Society of Testing and Materials, Philadelphia, PA, USA.

Barnthouse, L. (2008) On the use of mathematical models in ecological risk assessments: A response to Tannenbaum. *Integrated Environmental Assessment and Management* 4:4–5.

Brumbaugh, W.G., Ingersoll, C.G., Kemble, N.E., May, T.W. & Zajicek, J.L. (1994) Chemical characterization of sediments and pore water from the upper Clark Fork River and Milltown Reservoir, Montana. *Environmental Toxicology and Chemistry* 13:1971–1983.

Cohen, M.M. (2001) Frog decline, frog malformations, and a comparison of frog and human health. *American Journal of Medical Genetics* 104:101–109.

Dale, V.H., Biddinger, G.R., Newman, M.C. et al. (2008) Enhancing the ecological risk assessment process. *Integrated Environmental Assessment and Management* 4:306–313.

Dumont, J., Schultz, T.W., Buchanan, M. & Kao, G. (1983) Frog Embryo Teratogenisis Assay-Xenopus (FETAX) – A short-term assay applicable to complex environmental mixtures.

In: Waters, M.D., Sandhu, S.S., Lewtas-Husingh, J., Claxton, L.D., Chernoff, N. & Nesnow, S. (eds.) *Short-term Bioassays in the Analysis of Complex Environmental Mixtures III*. Plenum, New York, NY, pp. 393–405.

Forbes, V.E. & Callow, P. (2012) Promises and problems for the new paradigm for risk assessment and an alternative approach involving predictive systems models. *Environmental Toxicology and Chemistry* 31:2663–2671.

Giesy, J.P., Ludwig, J.P., &Tillitt, D.E. (1994) Deformities in birds of the Great Lakes Region; assigning causality. *Environmental Science and Technology* 28:128:128–135.

Lemly, A.D. (1999) Selenium impacts on fish: An insidious time bomb. *Human and Ecological Risk Assessment.* 5:1139–1151.

Peddicord R.K. & LaKind, J.S. (2000) Ecological and human health risks at an outdoore firing range. *Environmental Toxicology and Chemistry* 19:2602–2613.

Peddicord, R. (2010) Use of Peddicord & LaKind (2000) particle ingestion model in ecological risk assessment and site management. Prepared for Malcolm Pirnie, Inc., White Plains, NY.

Tannenbaum, L.V., Bazar, M., Hawkins, M.S. et al. (2003a) Rodent sperm analysis in field-based ecological risk assessment: pilot study at Ravenna army ammunition plant, Ravenna, Ohio. *Environmental Pollution* 123:21–29.

Tannenbaum, L.V., Johnson, M.S. & Bazar, M. (2003b) Application of the hazard quotient in remedial decisions: A comparison of human and ecological risk assessments. *Human and Ecological Risk Assessment* 9:387–401.

US EPA (1989) Risk Assessment Guidance for Superfund. Volume I: Human Health Evaluation Manual (Part A), Interim Final. US Environmental Protection Agency, Washington DC: EPA/540/1–89/002.

US FWS (2002) Ecological Risk Assessment for the Prime Hook Lead Pellet Site, Prime Hook National Wildlife Refuge, Milton, DE. US Fish and Wildlife Service.

Wirgin, I, Roy, N.K., Loftus, M., Chambers, R.C., Franks, D.G. & Hahn, M.E. (2011) Mechanistic basis of resistance to PCBs in Atlantic tomcod from the Hudson River. *Science* 331:1322–1325.

2 Facing the music: understanding what ERA is . . . and is not

Elementally, man's actions about the planet have often resulted in chemical additions to environmental media – soil, surface water, animal and plant tissue, etc. Health risk assessment as a process came about in response to the reality that environments are changed as a result of these additions, and even where the extent of the environmental change amounts to nothing more than the altered chemical composition of the receiving media. There are two altered chemical composition possibilities: either naturally occurring chemicals are found to be present in considerably greater concentration than was the case previously, or synthetic chemicals have been introduced. In either case, the health risk assessment question is the same. Since reasonably, exposure to chemicals can cause illness or even death, we are prompted to ask about the likelihood of the receptors who contact the media, developing toxicological outcomes that impinge on the quality of life. Further, we ostensibly want to know this because we recognize that receptors may not otherwise be able to circumvent their chemical exposures. A case in point is the US EPA's Superfund program, the program that effectively launched chemical health risk assessment as we know it. There, the specific risk assessment objective is to learn of the potential for health effects to arise in receptors, human or ecological, if there was no human intervention to remediate the condition of the chemically altered site.

ERA in the context of Superfund sites and the like has an established and agreed upon definition. It is the process that evaluates the likelihood that adverse ecological effects may occur or are occurring as a result of exposure to one or more stressors (US EPA 1992). This essential definition has been carried forward in unaltered fashion to later-developed guidance, as well it should (US EPA 1997,US EPA 1998); modification of something as basic as a discipline's definition can lead to confusion, and this would certainly be so for those affiliated in some way with the stewardship, assessment, or management of a contaminated parcel. A slightly different definition though, hails from the pivotal Superfund ERA guidance (US EPA 1989b) that precedes the US EPA's Framework for Ecological Risk Assessment

Alternative Ecological Risk Assessment: An Innovative Approach to Understanding Ecological Assessments for Contaminated Sites, First Edition. Lawrence V. Tannenbaum.

(US EPA 1992), and that curiously was published several months ahead of its human health companion volume (US EPA 1989a). Risk Assessment Guidance for Superfund-Volume II[1] defines ERA as "a qualitative and/or quantitative appraisal of the actual or potential effects of a hazardous waste site on plants and animals other than people and domesticated species." Regardless of which definition we might prefer, it should be clear to those who work in the ERA field that the definitions trigger numerous difficulties with far-reaching ramifications. It *should* be clear, but circumstantial evidence indicates otherwise. It's quite unlikely that the reader has come across discussion in the relevant literature, or heard directly from the mouth of any regulatory agency employee, that there exists no way to measure or assess ecological risk, this, the overarching difficulty that stems from our definitions. It is perhaps the ultimate irony; we cannot express the likelihood of adverse effects occurring in ecological (i.e., non-human) species, the very thing we as practitioners set about to do. This leaves us rather incapacitated. The situation is tantamount to defining transportation as the means of conveyance or travel from one place to another, yet there being no vehicles to accomplish such transfer on land, on water, or through the air. Consider that there is no such thing as medicine, other than what one might conjure up in the mind, if there are no drugs or potions to be applied in treating diseases. How the absence of a means to measure and express "the likelihood of" or "the potential for" effects that our ERA definitions highlight can go on unrecognized, or at the very least unacknowledged, can and should leave us stymied. Procedurally at some point amid every site's ERA negotiations, someone (preferably a regulator) should remark "Well, we must understand that none of what's been computed here tells us what the risk actually is. Let's not forget that no one can actually do such a thing." Such avowals never occur, however.

Leaving aside for now the curiosity of how there can be a generally unrecognized void in what we refer to as the ERA process, let us focus instead on a key distinction. Though this book will present alternative understandings and approaches for the science of what we term ERA, having identified the absence of a means of expressing ecological risk (i.e., the probability of one or more negative health effects arising), is *not* an example of one of those alternative understandings. The absence of a valid risk expression for ecological receptors is, rather, a fact. To see this we need to first observe that cancer as a toxicological endpoint is ecologically irrelevant; tumors, should they occur, do so in senescent animals only, well after animals are no longer reproductively contributing to a population. Vis-à-vis ecosystem health considerations of a contaminated site, it matters not that a bird, mammal, or fish develops cancer at such an advanced age. To fully appreciate that we have no ecological risk expression, we need to do one more thing. We need to

[1] This document has been superseded by US EPA 1997.

duly note that the mainstay of our ERAs, the HQ calculation, is not a risk measure (Tannenbaum et al. 2003; Kolluru 1996). The quintessential human health risk assessment guidance of the US EPA (US EPA 1989a) could not be more definitive about this. The reader should note that the underlining in the quotation that follows is taken directly from the guidance:

> The measure used to describe the potential for noncarcinogenic toxicity to occur in an individual is not expressed as the probability of an individual suffering an adverse effect. EPA does not at the present time use a probabilistic approach to estimating the potential for noncarcinogenic health effects. Instead, the potential for noncarcinogenic effects is evaluated by comparing an exposure level over a specified time period (e.g., lifetime) with a reference dose derived for a similar period.

It matters not that the above quotation comes from human health risk assessment guidance, for the essential HQ calculation is the same for both the human health and ecological risk assessment processes. HQs are simply dose comparisons performed discretely for each site chemical of potential concern that has the capacity to produce a systemic health effect in an exposed receptor. Absent a means of expressing health *risk* to ecological species from the chemical exposures they experience, we are plagued with the difficulty of naming the reports we generate. If nowhere between the covers of a thick, multi-volume report with a title along the lines of "Ecological Risk Assessment for the __________ Site", appear statements of the *probability* that a receptor (perhaps a hawk) will experience a toxicological endpoint of concern (perhaps delayed growth), the report is not a risk assessment. If on the other hand, we have learned that Red-tailed hawks utilizing a particular site have a 30% probability of exhibiting delayed growth, we *would be* expressing risk, and we would have technical matter about which to strategize when making remedial decisions. The nomenclature we apply to our would-be assessments is damaging on several scores. For other than the insightful, there is no realization that despite all efforts taken (in data collection, laboratory analysis, data reduction, chemical screening, statistical analysis, HQ and other computations, toxicity testing, etc.), the end user of the so-called ERA is not equipped with the knowledge of how probable it is that any of the terrestrial or aquatic receptors at a site will succumb to a health effect or impact. For the individuals comprising the workgroup attending to a given site, the damage takes another form, namely fully (albeit erroneously) believing that their comprehensive "ERA" report furnishes sufficient information to support a conclusion that a cleanup is necessary (with conclusions of that order far outnumbering conclusions to the contrary).

Inability to express ecological risk translates into more than would-be ERA reports being mislabeled, for all ERA *guidance* documents share a similar difficulty. There is no guidance document for ERA that communicates how to calculate or express the probability estimates that the science (we would hope) longs for. The damage done here is in misguiding the guidance user, i.e., leading him or her to think that with instruction provided on such things as formulating a proper problem statement, identification of a receptors-of-concern list, selection of a suitable reference location, and conducting a phased assessment, the answer we seek will follow. Although it is true that the information the available guidance enables us to assemble leads to constructive field sampling events, appropriate analytical reporting, computed HQs, and worthy weight-of-evidence discussion, these documents do not get us to a supported decision point. It is of course, disheartening to witness a continued inattention to the reality of plagued and mislabeled guidance. Perhaps the classic case in point relevant to this discussion is the EPA's guidance for conducting probabilistic risk assessment (PRA) intended to address the needs of both HHRA and ERA (US EPA 2001). Early on in this document, we come across a fatal flaw; PRA, we learn, is a risk assessment that uses probability distributions to characterize variability or uncertainty in risk estimates. As discussed just above, the lone numerical expression in ERA, the unitless HQ ratio, is not risk estimate. Absent a risk estimate, PRA for ERA does not exist. The PRA guidance raises false hopes with its title, but bears no practical assistance to the crippled ERA field.

That a need for generating guidance that breaks free of the status quo fails to register on the proverbial radar screen, is demonstrated by a recent US EPA initiative (US EPA 2010). In recognition of the agency having produced so much guidance in support of ERA over the years, an effort was undertaken to synthesize the information contained in more than 50 distinct documents. The culmination of this undertaking is a report appendix consisting of a multi-page table that lists the varied guidance documents to emerge from the agency over a period of two and a-half decades. Through a matrix format, the table indicates which of some 39 aspects or criteria of the larger ERA landscape are mentioned or discussed to some extent. The ERA elements of the matrix include "document scope" (e.g., science policy, planning and scoping), the ERA process components of problem formulation and analysis (e.g., assessment endpoints, effects analysis), "related components" (data quality objectives, QA/QC), "corollary issues" (e.g., stakeholder involvement, benefit services, etc.), "regulatory framework/program" (e.g., TSCA, pesticides, etc.), "level/scale" (e.g., population, landscape), "scope of guidance" (e.g., media, receptor, etc.), and "type of assessment" (e.g., causal, predictive, etc.). Though this matrixed compendium will direct the user to documents that may be of interest and that may otherwise be hard to locate, it is imperative to recognize that the initiative does not embark on any new ERA territory. Though a well-intended exercise, the effort is nothing more than a guidance recycling effort, and as such it seemingly only serves to reinforce the (incorrect) message that the information on how to truly assess ecological risk exists and is available for the taking.

A last ramification of an absent means of calculating or expressing ecological risk, bears on those employed in this topical area. One who considers himself to be an "ecological risk assessor" is not so. He or she may be a reviewer of documents that relate to the assessment of ecological effects, one who calculates HQs, or one who compares contaminant concentrations in environmental media to tabular values of reputedly safe-level or effect-level screening benchmarks, but this person cannot *de facto* be an ecological risk assessor. In and of itself this mislabeling is innocuous. Individuals hired under such a title can collect a legitimate salary and need not fear being audited because they portray themselves with the ability to articulate the probabilities of ecological receptors developing and displaying health impacts when in fact they cannot render such a service. The *only* harm done with this job titling is the self-deception and artificial empowerment it fosters. So-called ecological risk assessors who elect to attend a conference or workshop on ERA may find themselves sitting in a room with others who share such a designation, but none are working towards a goal of crafting a means by which the probabilities mentioned many times in the foregoing, can be expressed.

If expressions of ecological risk are not being generated, what then are ERAs reporting? It would appear that identifying settings and circumstances where unacceptable risk could be present is what ERA practitioners have either consciously or subconsciously put forth in risk's stead. This substitution, not surprisingly, introduces a new set of difficulties. Risk factors are present everywhere, and risk assessors know that the notion of "zero risk" doesn't exist in any sphere. In replacing what remains to be the elusive computation and expression of risk, with an itemization of factors that could contribute to risk, the science reduces to guilt-by-association. In other words, if there are chemical stressors present onsite, it is concluded that ecological receptors there must be at an unacceptable risk level. While the reader may not wish to acknowledge that ERA practitioners so conclude, every instance of a recommendation to clean up a site based only on the occurrence of one or more HQs above 1.0, unmistakably proves the point. Whose task is it though, to determine that intervention (most often in the form of site remediation) is needed to minimize or abate supposed unacceptable risk, and what reliable and technically defensible criteria are to be utilized in making such a determination? The following account of an actual site visit will solidify the point.

The site is a relatively narrow perennial creek (4–6 feet across) that bisects a formerly used small arms range on a military installation, and that under base flow conditions is approximately 0.5–1.0 foot deep. The creek's substrate is comprised mostly of boulders and cobbles with a few small depositional areas of sand substrate. Steep embankments approximately 4 feet high on the creek's south side acted as earthen berms when the range was operational, and erosional processes over the years have deposited M-16 and .45 calibre bullets and bullet fragments into the creek channel. For most of the approximate 3200 feet of creek that flows through the range, bullet slugs are readily apparent, glimmering in the water on the sunny day when the site visit occurs. Attending the visit are installation personnel (some of whom work in the post's environmental office), civilian scientists, risk assessors employed by the military, and state and federal regulators. A few individuals squat down on cobble to pull out a few slugs by hand. Several individuals are heard to remark that the slugs will need to be removed, and several others nod in agreement. There is no question that the creek would more look the part if its bed was swept of the slugs thereby removing evident point sources of lead, but what was the justification for the remark that slugs be removed? Since ERAs are not driven by a directive to ensure that measures are taken to enhance or restore a contaminated site's *appearance*, a proper answer is sought. There had been no demonstration of unacceptable risk for ecological receptors that might utilize the creek. In fact, the list of such ecological receptors is rather limited. Fish have never been observed in the creek despite water flow being maintained at all times, to include periods of drought. The receptor list appears to be limited to only the aquatic insects that develop there. Of course, sites are not remediated to protect insects, but their potential role in serving as diet items for other receptors in the locality cannot be ignored either. Here though, we again fail to discover a valid ERA contaminant exposure pathway of concern. As per the site-specific reporting (Shaw 2010), the creek provides only low-quality foraging habitat for the Federally-listed endangered gray bat (*Myotis grisescens*), because the bat's requirement of continuous cover while traveling to and from its foraging habitats is not supplied by the creek.

High metal concentrations in the water, traceable to the submerged slugs, could not have prompted individuals to opine about the slugs removal, for no one on the site visit knew that such concentrations might have manifested themselves. If the remark reflected a concern over the protection of benthic infauna (e.g., critical life stages of insects and other crustaceans), those calling for remediation hadn't reviewed any data on benthic assemblages for the creek and a comparable aquatic habitat reference point. In fact, such a benthic assemblage comparison for the creek hadn't been conducted.

For the above account, no one calling for the creek remediation had the information needed to constitute a valid justification for it, nor had they considered that the stretch of affected creek had borne the cache of bullet slugs for the years of small arms range operation as well as the decade following range closure. What mattered and what gave voice to the motion put forth to remediate the creek were two things: the creek not appearing as it should in nature, and a tenuously constructed chemical exposure scenario whereby dissolving bullet slugs undoubtedly release lead to the creek, with the lead additions in turn causing ecological receptor impacts. Neither of these sentiments is defensible. As already mentioned, an altered site appearance is not a trigger for remediation, and certainly not when operating within a risk-based program. It is possible that ecological receptors utilizing a creek laden with bullet slugs act differently or are incapacitated in some way because of the visual cues or direct toxicity they experience. In the absence of a well-designed study that could prove to a respectable scientific standard that creek receptors are disaffected because of their changed environment, calling for a remedial effort is premature, reflecting only an emotional tie to the situation at hand. Undoubtedly certain site visitors were bothered by the prospect of leaving a seeming contamination source in place. The simplistic and loosely assembled exposure scenario has two trappings. First it amounts only to conjecture, with stakeholders (only) imagining what will happen. For ERA purposes though, decisions to invoke remedial action should not be based on conjecture. Secondly, the exposure scenario reflects bias in the form of assuming that human toxicological responses (in this case, to lead) directly apply to the ecological receptors.

In the main, with the absence of a means to calculate and express ecological risk, ERA has been corrupted largely by a misunderstanding and misapplication of routinely invoked chemical screening exercises. The aforementioned bullet slug case provides a first, albeit simplistic, example of this. A comparison was conducted of the described creek and a mind's-eye image of a comparable creek without the slug additions. Since the former differed in appearance from the norm, it was concluded that it needed repair, this while no one could articulate what element of the creek system was experiencing unhealthful effects or was in need of additional protection. For many, the jury is simply "out" where an anthropogenic or xenobiotic chemical is discovered; anything encountered in the site media that is out of the ordinary is perceived to be a problem. The discovery of trichloroethylene (TCE) or PCBs in site soil or sediment amounts to indisputable evidence of both a prior human presence and human wrongdoing, but it does not summarily prove that ecological receptors interacting with these media are at risk. To remark (as in the case of the bullet slugs) that the soils contaminated with TCE or PCB will have to be remediated, as is often articulated in face-to-face stakeholder meetings, is to abuse the skeletal process that we do have. Passing judgment on anthropogenics and xenobiotics is effectively concluding that chemical presence *alone* is a problem, one that poses unacceptable risk. It is tantamount to saying that since the chemical shouldn't be here at all, it must be bad, and because it is bad, it must be removed. If mere chemical presence is to decide that a cleanup must follow, why then have regulatory bodies dabbled in risk assessment at all over the years, and produced guidance to address contamination that we encounter (putting aside for now that at best, the guidance is meager)? To have chemical presence alone cast the die advising or insisting that cleanups ensue, is to make short-shrift of what our field data collection efforts could be. Seemingly one would want to know how extensive ecological receptor exposures are, how long a given site has borne its chemical profile, and whether or not the chemical profile has changed over time. It is entirely possible that a xenobiotic load has chemically degraded with the passage of time or has become bound to the soil or sediment moiety in such a way that a toxicological response can barely be elicited. Of course, our bullet slug scenario and many other real-world contaminated site arrangements could well avoid getting tripped up in guilt-by-association thinking, were there to be a proliferation of biomarker science. Up to this point, ERA has not championed this approach, whereby one could know that a receptor's health is offset based on the presence or appearance of a measurable characteristic serving as a condition or disease indicator.

The misapplication of ERA screening efforts, in any of the varied forms they can assume, has left a devastating impact on ERA, although many probably fail to realize this. Screening efforts have been empowered by the rank and file; in a very real way, having screened has come to be synonymous with having assessed risk. Screening of course, was never meant to be a surrogate for risk assessment, and screening's true place should be only as a forerunner to it, with screening outcomes either indicating that risk assessment tasks should proceed, or that risk assessment is not needed. One can only speculate how screening efforts rose to such power, and it would serve no purpose to set out to decidedly uncover how the field has come to be as reliant on the HQ as it is. It would be a purposeless exercise because the ERA establishment has never expressed an interest in adopting a new paradigm free of HQs, and likely never will. It might be worthwhile though, to mention two possibilities that can account for the HQ having come to wield the power that it does. First, it could be that ERA's early craftsmen lost sight of the actual mathematical meaning of the HQ. Alternatively the early ERA craftsmen may well have been aware of the HQ's limitations, and at the same time come to the realization that assembling an assessment scheme notably more advanced than the HQ, was simply too challenging a prospect to take on. Just how would one go about demonstrating the *probability* that a site receptor might develop a toxicological effect? There has never been an answer to this question, and later on we will see why an answer is not needed. For our purposes it is important to note that even the HHRA process at best only does a marginal job in expressing risk, because it can do so only for the cancer endpoint (and not for systemic effects), one of two types of toxicological manifestations that concern us. If the relevance of a cancer endpoint is absent within ERA, as has been mentioned previously, any toxicological data we might have at our disposal cannot be used to generate probability-of-toxicological outcome information.

How could an ERA practitioner think that either the HQ or its relative, the ecological effects quotient (EEQ), are risk expressions in the first place, let alone defensible ones? Before we allow the matter of these expressions being devoid of units to interfere, let us consider the toxicological source material of an HQ. Invariably the studies that support the toxicity reference values (TRVs) that form the HQ denominator reflect no less than a standard set of some six or seven formidable departures from the case of the specific receptor in the field that one intends to assess. Remarkably, a number of these are almost never mentioned in the uncertainty sections of ERAs. As we review most of the departures here, the reader should note that they are also discussed in a toxicity assessment context in Chapter 4.

1 We begin with the species difference. Since all studies supporting TRVs occur in the laboratory and therefore employ conventional rodents and just a few other species, this departure is, of course, unavoidable. As closely related in appearance and physiology to the would-be-assessed field species as the surrogate lab species might be, the two are not the same. Our best chance of a very close alignment in dose–response would seemingly involve a conventional laboratory mouse (e.g., *Mus* sp.) or rat species (e.g., *Rattus* spp.) standing in for the meadow vole (*Microtus pennsylvanicus*), the pocket mouse (*Chaetodipus* spp.), or the Hispid cotton rat (*Sigmodon hispidus*) of the contaminated site.[2] The only lingering matter here is our need to come to grips with the utility of rodent assessments overall. Are assessments for site rodents needed in the first place? While we may have never come across a terrestrial ERA that did not evaluate at least one rodent species or a close relative, it is a reality that we do not clean up sites so as to afford protection to rodents. In effect then, although laboratory-rodent-derived TRVs may be salvageable in our discussion (of TRV applicability) because they do support assessments for their field-residing phylogenetic relatives, we really haven't gained much if any ground. How comfortable are we with using dose–response curve information of *Mus musculus* in (HQ-based) assessments for squirrel, fox, or mink? If species difference was our only departure, there might be merit to the calculated HQ. The reader should note that this review exercise of toxicological source material is not simply an opportunity to critique ecological HQs. HHRAs really fare no better as they too almost exclusively rely on rodent dose–response. There are however, two saving graces to HHRA. First, while we may be erring in applying rodent-derived safe-level and effect-level doses, we are at least doing so in a predictable and highly consistent manner; our extrapolations in HHRA are always of the "rodent-to-human" genre. Secondly, in all aspects of medicine and public health, we aim to be extraordinarily conservative, and

[2] The ever-popular shrew, as in the Short-tailed shrew, *Blarina brevicauda* of terrestrial ERAs belongs to the mammalian order of *Insectivora* and not *Rodentia*.

we are quite upfront about this. Where human health is being considered, the slightest hint of a potential health effect is deemed unacceptable. A case in point and perhaps a best example is that of reduced sperm counts in laboratory test animals. As dealt with in much greater depth in Chapter 9, we are fortunate to know the degree to which sperm counts need to drop in order to pose a reproductive threat (e.g., for there to be fewer young born). In pharmaceutical testing for the development of new products, were there to be a statistically significant count reduction in treated animals relative to controls, however slight the difference, the drug would unquestionably need to be reworked. Why the pharmaceutical team would return to the drawing board when the well-established sperm count reduction threshold-for-effect wasn't approached even by a long shot, reflects the imposed ethically based priority we place on preserving our own species.

Beyond rodent-for-rodent exchanges where there might be similarity in dose–response, we are otherwise far afield. We should not suspect that adjustments to dose–response information for such things as body weight, metabolic rate, and a host of other physiological parameters will correct for inherent species differences within a phylogenetic class or order. On the avian side, we cannot summarily claim dose–response information of Japanese quail or chicken to be utilitarian when it is a robin, vireo, pheasant, or Northern harrier that is to be assessed. In truth, we are not interested in knowing how a laboratory-reared rodent or avian species of common choice responds to its dosing, and certainly when there are so many other additional departures from the case of the potentially affected animal at the contaminated site.

2 The established laboratory testing scheme routinely controls the test animal's immediate environment. As mainstays, the room temperature and lighting are rigidly set, but we know that in the wild the animal to be assessed experiences anything but fixed (i.e., constant) temperature and lighting in the course of a day, much less a week, a month, or a season. A laboratory test animal is limited to the area and the contours of its cage or enclosure, whereas the animal in the wild has vastly more area to explore and utilize, and all in the context of an immeasurably more varied and non-uniform environment. It is reasonable to expect that these differences influence the manifestations of exposures to toxic compounds. Although there is most definitely a need to conduct dose–response studies under highly controlled conditions, the departures from the real-world exposure case add appreciably to our uncertainty regarding the potential effects of chemical exposure.

3 The mode of chemical administration in dose–response studies frequently does not coincide with the manner in which ecological receptors in the wild incur their chemical uptakes. Administering a chemical via a syringe for example, as in an intraperitoneal injection, ensures both the full delivery of a chemical

dose and the chemical being delivered to the desirable location of the body, but these features come at the expense of the delivery system being most unnatural. Keeping with this example, the content of the syringe must also bear up to scrutiny. We cannot summarily assume that a chemical dissolved into a liquid vehicle for efficient dosing (e.g., corn oil) is toxicologically received as it would be by a receptor of concern in its natural habitat. In the span of time it takes to inject or gastrically intubate a laboratory-reared animal, a receptor in the wild might not consume an equivalent chemical dose, in terms of potency and volume, to that administered in a controlled experiment. Further, a chemical that exists as a solid (e.g., a powder), and that would be consumed as such in the wild, is undeniably a different entity when dissolved into water or some other liquid. The chemical's metabolism might be reduced or augmented because of the change in physical state, and the time until the chemical first appears in the bloodstream will undoubtedly be influenced by the difference in physical state as well. While the arrangements of dose–response protocols are quite understandable, we should be concerned about the inattention given to the uncertainties associated with the TRVs that are constructed in the aftermath of the conducted studies. From the toxicologist's perspective (be it the one conducting the dose–response study or the one reading about it) there was a great need to standardize the study design, striving for great uniformity with respect to animals receiving their chemical-in-diet treatments. Seemingly the quest for such standardization and uniformity overshadowed the toxicologist's concern with mimicking the chemical exposures that occur in the wild. As we will learn, the overwhelming majority of dose–response studies that support TRV development preceded ERA, and so there is no blame to assign to the toxicologist. If there is any blame at all to parcel out, it should be directed to those who derive TRVs and apply them, all the while failing to acknowledge (in ERA reports and the like) the departures from the real-world condition inherent in the testing.

4 It is a rare occurrence that a site contaminant's specific chemical form is identified in ERA doings. While there may be legitimate reasons for not speciating chemicals, such as excessive costs involved in securing this level of refinement, developed TRVs available for use come at a cost as well. Thus the integrity of the HQs to result from TRV application is threatened because the TRV may only be reporting on the toxicological responses to variants of the chemicals that are actually present in the environmental media of interest to us.

5 Unless there is but a solitary contaminant (of concern) in the medium to which an ecological receptor is exposed, receptors are faced with a fully integrated suite of contaminants in an ongoing fashion. We should then, seek to know what the collective toll of the contaminants is on receptor (or better,

population) health. The HQ construct, however, is not capable of speaking to more than one chemical at a time, evident in the bullet-point reporting style of ERA summary or conclusion sections. We learn that Chemical A does not pose a hazard (or a potential for risk), while Chemical B does, etc., etc. Acknowledging our inability to craft more constructive statements is, of course at the very least, the correct thing to do. Such hardly occurs though, but even if the avowals were made, we need to assess where the anemic reporting leaves us. It is unfortunate enough that non-risk expressions are being generated, but we are more than hard-pressed to find utility at all to HQ expressions when they fail to consider the potential synergistic effects of multiple chemicals. Plausibly, the effects of some chemicals are muted if not cancelled out entirely by others. As the challenge to distinctly point to sites that have something wrong with them continues to not be met, at the very least, we probably need to ascribe the apparent chemical immunity to the competing (negating) actions of chemical stressors.

6 As if our list wasn't long enough – of ways in which TRV-supporting laboratory studies substantially depart from the exposure case of the chemically exposed animals in the field – we will add one more (for now) that is never acknowledged in the uncertainty sections of ERAs. For obvious reasons we are primed, of course, to select chronic duration studies as opposed to acute duration studies. While this makes for a good start (although our efforts are again aimed only at HQ computation), there is just so much that we can laud ourselves for with regard to this study choice. While a 3-month or 6-month study is superior to a 2-week or 4-week one, the fact that our selected studies all occur within the lives of test cohorts is consistently overlooked. Such studies are not what we really need to know about, certainly not when our focus is on sites that are decades old and where multiple generations (perhaps hundreds) have already dealt with an onslaught of contamination. It is true that some receptors, site immigrants, will encounter site contamination anew, but it is unlikely that new site immigrants describe the dominant make-up of the forms that populate Superfund sites and the like. Realistically, we should not anticipate a research orientation shift to suddenly overtake us, whereby multiple-generation exposure studies handily arrive to replace the one-generation studies with which we are so familiar. As with the previous departure type reviewed, the reader should appreciate not only the inattention to the one-generation study/multiple-generation study discrepancy in uncertainty section reporting, but the consequences of our utilizing one-generation studies in HQ computation. For the overwhelming majority of site receptors that are not new, but have instead either lived for years in a contaminated environment or have been repeatedly exposed to site contaminants through maternal transfer events, we should anticipate a toxicity response capacity

that is far different; unless sites of interest are devoid of biota, the receptors can (easily) tolerate the toxins around them. Before applying toxicity response information of a kind that we clearly know not to be relevant to the contaminated site condition, we should ask if our sites of interest support biota. If they do, and if in particular they support the very same toxicity test species, it is then clear that HQs aren't needed. If unacceptable HQs were computed, yet sites were observed to be supporting the species being evaluated through HQ computation, would we conclude that the species observed *in situ* weren't really there, but were figments of our imagination?

The finger-pointing at the HQ construct is well-deserved for two reasons: the shortcomings reviewed above are all real, and we elect (nevertheless) to employ the HQ as the be-all and end-all of our assessments. We will conclude this chapter's review of what ERA is and is not, with a return to the concept of screening, specifically with a HHRA–ERA comparison.

Screening in HHRA amounts to reviewing the chemicals of a site to determine which ones are appropriate for being further considered, or as the parlance has it, "to be carried through the risk assessment". The outcome of screening in HHRA is a simple list of chemical names and nothing more. Assuming one or more chemicals were retained through the screening process, the risk assessment (the HHRA) will report on the capacity of the chemicals (singly or in combination) to increase the likelihood of a toxicological endpoint to be reached. In ERA, screening and what is termed the "risk assessment" are identical operations, with the HQ construct being used each time (Tannenbaum et al. 2003). HQs of each operation differ only in magnitude; with the conservatism of the initial HQs relaxed, the latter HQs will necessarily be smaller numbers. Importantly, if the HQ is a screen (which it is), it is so in every case and even if there were to be multiple adjustments (i.e., more than two tries at its computation) to the chemical intake expression components (e.g., percentage of the diet made up of a certain foodstuff, body weight of the modeled receptor, etc.). For all of our efforts, what we get for our money with an ERA is a unitless expression that supposedly articulates the multiple (possibly a fractional one) of what is thought to be a safe-level or effect-level dose. ERA is not capable of anything more than (crudely) screening chemicals with the intent of identifying which ones might be problematic for site receptors.

References

Kolluru, R. (1996) Health risk assessment: principles and practices. In: Kolluru, R., Bartell, S., Pitblado, R. & Stricoff, R.S. (eds.) *Risk Assessment and Management Handbook.* McGraw-Hill, New York, NY, pp. 10.3–10.59.

Shaw (2010) Baseline ecological risk assessment for Iron Mountain Road and Bains Gap Ranges, Appendix J. Shaw Environmental & Infrastructure Group, Fort McClellan, Calhoun County, Alabama.

Tannenbaum, L.V., Johnson, M.S., Bazar, M. (2003) Application of the hazard quotient in remedial decisions: A comparison of human and ecological risk assessments. *Human and Ecological Risk Assessment* 9:387–401.

US EPA (1989a) Risk assessment guidance for Superfund. Volume I: Human health evaluation manual (Part A), Interim Final. Washington DC: US Environmental Protection Agency. EPA/540/1–89/002. Washington DC: US Environmental Protection Agency.

US EPA (1989b) Risk assessment guidance for Superfund. Volume II, Environmental Evaluation Manual, Interim Final. EPA/540–1–89/001. Washington DC: US Environmental Protection Agency.

US EPA (1992) Framework for Ecological Risk Assessment. Washington DC: Risk Assessment Forum. EPA/630/R-92/001. Washington DC: US Environmental Protection Agency.

US EPA (1997) Ecological Risk Assessment Guidance for Superfund: Process for Designing and Conducting Ecological Risk Assessments, Interim Final. EPA/540-R-97–006. Washington DC: US Environmental Protection Agency.

US EPA (1998) Guidelines for Ecological Risk Assessment. Washington DC: Risk Assessment Forum. EPA/630/R-85/002F. Washington DC: US Environmental Protection Agency.

US EPA (2001) Risk Assessment Guidance for Superfund: Volume III – Part A, Process for Conducting Probabilistic Risk Assessment. Office of Emergency and Remedial Response, EPA 540-R-02–002. Washington DC: US Environmental Protection Agency.

US EPA (2010) Integrating Ecological Assessment and Decision-Making at EPA: A Path Forward; Results of a Colloquium in Response to Science Advisory Board and National Research Council Recommendations. Washington DC: Risk Assessment Forum. EPA/100/R-10/004. Washington DC: US Environmental Protection Agency.

3 Alternative exposure assessment

In day-to-day ERA application, exposure assessment is a rather scripted affair. Right or wrong, a receptors-of-concern list will always be compiled, for HQ computations must happen. Presumably then, in each and every site instance, one or more receptors are chemically exposed to make it worth our while to compute the HQs. An unwillingness to ever dispense with the HQ calculation is one example of the scripted nature of exposure assessments – receptors might not be sufficiently exposed to legitimize HQ computation, but it is assumed otherwise so that the process can play out. Exposure assessments are of course, scripted in other ways. Thus, only one contaminant uptake route (ingestion) is ever manipulated in the case of mammals and birds, while the existence of other routes is, at best, acknowledged. (For fish and other aquatic species, owing to a lacking contaminant uptake HQ computation process, assessments take the form of comparing contaminant concentrations in the water column against tabular screening concentrations that are supposedly protective.) With regard to the "other contaminant uptake routes", the scripting extends to the supply of hefty canned language explaining why these are not operative or viable. But what if they are? More recently, scripting in exposure assessment has taken on a very obvious visage, namely the appointing to the receptors-of-concern list, at least one animal for each feeding guild or strategy. Seemingly an ERA would be shirking its responsibilities or doing an incomplete job if it didn't consider a mammalian herbivore, carnivore, and omnivore. There is often the sentiment as well, that the job done is less than complete unless it considers both a small and large mammalian herbivore, carnivore, and omnivore. On the avian side, we have come to expect to see profiled for evaluation, a granivore, a vermivore, and an invertivore (and maybe too, a piscivore). But are we so certain that we reliably have these many forms present at our sites? Scripting in exposure assessment can also take the form of elaborate discussion centered on (chemical) bioavailability. For the present chapter, bioavailability will serve as the starting point for an alternative understanding of exposure assessment.

Bioavailability, an important contaminant characteristic that influences the degree of chemical–receptor interaction, remains an ongoing high interest area

Alternative Ecological Risk Assessment: An Innovative Approach to Understanding Ecological Assessments for Contaminated Sites, First Edition. Lawrence V. Tannenbaum.

in ERA. To say that large numbers of ERA practitioners are fascinated by the role played by bioavailability is to seriously understate the case. Attesting to the popularity of the topic are workshops and professional society conferences that convene only to explore aspects of the phenomenon, and a vast growing literature of published papers particularly reporting on sediment ecology. Suffice it to say that those who are wont to summarily conclude that ecological problems exist based only on the presence of chemicals in the environment, as in the case of the creek full of bullet slugs profiled in Chapter 2, do not constitute the entire lot of ERA practitioners. There is a deep appreciation for the reality that the true chemical exposures to plant and animal receptors with which to be concerned occur at those interfaces where the site chemicals could traverse the cell walls and cell membranes they are targeting. There is thus a recognition that mere chemical presence in an environmental medium tells far less than the whole story. The heart of the matter instead lies in accounting for that component of the total chemical load present in the medium that is available to be taken up by the receptor. Given the highly complex nature of environmental matrices, due consideration must be given to chemical binding affinities and chemical fate and transport processes. After acknowledging the chemical partitioning that occurs, endeavoring to estimate, if not directly measure, how much of a chemical is available to physically enter the organism (through the mouth, lungs, integument, roots, stems, leaves, etc.) is recognized to be a worthwhile undertaking. Sequentially the next undertakings, drawing on applied pharmacokinetics, seek to estimate or measure how much of the chemical will enter the cell, how much of the chemical will elicit a toxicological response, and finally, how severe a toxicological response is likely to ensue.

Perhaps there is another and far more intrinsic way to understand bioavailability, and perhaps ERA practitioners have jumped the gun in their approach to quantifying chemical uptake by contaminated-site receptors. An alternative approach to bioavailability considers the opportunities for site receptors, through their species-unique spatial movements, to physically encounter chemicals altogether. Where the frequencies and/or the durations of such encounters are low or minimal, efforts to quantify chemical intake such as those described earlier are seen to be misguided and completely unnecessary. Why bother with characterizing site soil or sediment in a highly detailed fashion to understand the capacities of these media to sequester chemicals, if the receptors we wish to protect are hardly present at that soil or sediment in the first place? We are speaking here, of course, where a receptor's natural placement in the environment as opposed to any habitat disruption due to humans that may have occurred, accounts for low receptor numbers. Additionally, the reader should note that it is not by any means standard ERA fare to characterize contaminated site habitats (overall), let alone to do so in a sufficiently sophisticated manner to allow for findings of reduced receptor numbers to be ascribed to compromised habitat. Drawing again on the case of a terrestrial site setting, why conduct a certain specialized suite of earthworm toxicity tests if there are insufficienct vermivorous receptors utilizing the site to give us cause for concern. Bioavailability then needs to be redefined for ERA purposes. A chemical is bioavailable in a meaningful way only if a site's receptors of concern can contact the medium containing the chemical to a point where it is reasonable to expect that a toxicological response of import could be elicited.[1] Additionally, a chemical is bioavailable for ERA needs only if there is a sufficiency of receptors contacting the contaminated medium to constitute a legitimate concern. In the quest for gaining a practical understanding of bioavailability, a shift away from the present-day cell interface focus is sorely needed. Procedurally we would *not* commence with an exploration into the mechanisms involved in taking up a chemical from the environment to be subsequently metabolized; we would instead *grapple with* (but not really, see later explanation) the matter of whether or not we are faced with a chemically contaminated site deserving of our attention because it seemingly poses a threat to the biota that reside there. Where the ERA practitioner might have gone so far as to relegate bioavailability to a category all of its own, and to commit to investigating the related phenomena of bioconcentration, bioaccumulation, and biomagnification, then bioavailability as described here is an ERA element safely parked within the long-standing ERA process step of Exposure Assessment (US EPA 1997; US EPA 1998). In this alternative view, where one could have asked about how much of a chemical might be expected to pose toxic action once inside a cell, we discover a reframed question – certainly in the case of the terrestrial site – namely, "How large is your contaminated site of interest?" As we shall learn, in many instances there is abundant, readily available information to demonstrate that our sites are often not large enough to matter. The above usage of "grappling" then, with regard to the task of deciding if a site genuinely presents a problem to us, came only to dramatize the distinction in approaches to addressing bioavailability. Understanding chemical uptake in terms of chemical partitioning in the affected matrix, and in terms of cellular and subcellular toxicity, are huge undertakings that demand great expenditures of time. In stark contrast, just a minimal investment of time will reveal that we haven't many sites to worry about from a vantage point of the spatial relevance/irrelevance of receptors. If there could be any *challenge* at all connoted by the word "grappling" here, it would apply to ERA practitioners who have jumped onto the bandwagon that hails cell membrane-based bioavailability as "the" topical area to map out. Entrenched in such thinking, this population could be faced with the *challenges* of absorbing a new and much more basic definition of bioavailability, expanding the list of tasks

[1] Efforts to apply or adapt existing ERA frameworks to consider singular chemical exposure events of certain receptors may be well intended, but in truth, we do not know of the existence of "fly by and die sites." No Superfund-type site has ever been described as being toxic to the point that death is anticipated pursuant to a receptor encroaching on a contaminated site just once.

that fall under Exposure Assessment, and resigning themselves to the reality that there may not be many sites worthy of our attention.

Our first foray into embracing exposure assessment and bioavailability in a more practical way, reduced the question of "How much of a chemical is free to enter the cells of a receptor to pose toxic action?", to the question of "How large is the contaminated site of interest?" As explained in the earlier chapters, sites where conventional ERAs are applied are painfully small relative to the density and spatial movements of meaningful ecological receptors. In acknowledging this truth, we can reduce the exposure assessment/bioavailability question two more times, first generically, and then site-specifically. We should first ask "How many species representatives are we concerned about?" After deciding this question, which is applicable to all sites, we then need to answer the (site-specific) question: "How many representatives of species x are present at our specific site of interest?" If due to natural means a site of interest doesn't supply or support the requisite animal numbers for true concern, an ERA isn't needed. There is nothing crude about the question sequence, one that is sincerely designed to determine whether or not a site of interest is marked by one or more completed or sufficiently completed exposure pathways, which are indispensible elements for any health risk assessment. Several qualifications are required at this juncture, however, and these are addressed in the following paragraphs.

- The generic question is quite honestly introducing the concept – for ERA purposes, that is – of expendable animals. To a vested-interest party whose primary or only focus is other than risk assessment and instead squarely placed on wildlife protection, the concept would undoubtedly be taken to be abhorrent. It is critical to understand that promoting a concept of expendable animals would certainly not be done to deliberately anger vested-interest parties. Equally so, it is not ERA's place or responsibility to placate natural resource trustees or their designees who might find themselves disgruntled by what they might see as a cavalier approach to the consideration of site receptors. Importantly, there are options available to vested-interest parties for constructively dealing with their degree of upset, and one of these is to pursue bringing natural resource damage claims, where they are able to.

  It is important to recall here that natural resource trustees (e.g., the US Department of the Interior) or their designees (e.g., the US Fish and Wildlife Service; FWS) are not part-and-parcel entities of the Superfund ERA process.[2] Their connection to the process extends only to being apprised of the schedule of meetings and other communications concerning the management of

[2] It is most curious that on multiple occasions the US EPA, due to staffing shortages, etc., has allowed some of its Superfund program Biological Technical Assistance Groups (BTAGS) to be led by FWS employees. With FWS espousing scientific approaches and assessment principles that are far removed from EPA doctrine, it is no wonder that site-specific proceedings are commonly drawn-out affairs, and that a uniform approach to ecological site management is not found within the ranks of the EPA.

contaminated sites, and to being regular recipients of site-specific documents generated as part of the Comprehensive Environmental Response, Compensation and Liability Act (CERCLA) Remedial Investigation/Feasibility Study (RI/FS) process (US EPA 1992b), a point often forgotten. An agency such as FWS has its own agenda to follow, and one that doesn't neatly mesh with the ERA process, which is evident in that the loss of even one species representative (e.g., one squirrel) is unacceptable if something could be done to prevent it. We can well understand why such parties would object to a question sequence that espouses the concept of defining a site's receptors-of-concern list on the basis of the number of species representatives expected to be present at a site. To such parties, the premise smacks of knowingly disregarding investigations concerning health effects that might accrue to species that they work so hard at protecting. Yet, for the betterment of ERA science, we need to raise these never-before-asked questions and to be willing to discover that it may be considerably more than one species representative that can be deliberately ignored or overlooked.

- The answer to the allowable quota question (regarding species representatives that we need not concern ourselves with, although these animals are undoubtedly chemically exposed) is not "one". Initially we might think to disagree with this because in HHRA, a singular reasonably site-placed receptor legitimizes a risk assessment undertaking. By way of example, the early years of the Superfund era often considered the actual case of the lone resident, who unlike his township neighbors with contaminated groundwater wells on their properties, refused to cease using his private well water for drinking and showering. It was established long ago in ERA that it is the *population*, as opposed to the individual, that is to be protected. En route to answering the allowable quota question, we must recognize then, that the absolute lower limit of concern cannot be any less than two species representatives, this in order to maintain the capability of the species representatives to continue to produce viable young.
- Asking about how many species representatives really concern us, is not meant to suggest that we are playing G-d, in the sense of our deciding who shall live and who shall die (not that we seem to know of any contaminated sites where animals die as a result of their contaminant exposures anyhow). The question rather is formulated in order to arrive at a common sense management approach to conducting ERA, and the question can be made more palatable by rephrasing it, or more precisely front-loading it or back-loading it vis-à-vis the CERCLA ERA process (Tannenbaum 2005b). The front-loaded question recognizes all that goes into an ERA becoming a reality: convening a kick-off meeting (of site personnel, federal and state regulatory agency representatives, risk assessors, and other technical support individuals), bringing one or more contractors on board, having a work plan produced and ultimately approved by all parties,

deploying to the field for data collection, having the data analyzed by a qualified laboratory, and producing a multi-volume RI that also needs to be approved by the respective parties. If it was known that a contaminated property supported three raccoons or two long-tailed weasels, would it make sense to launch into the process as it has been described here, with all of its inherent costs and resource investments? If we had somehow shown unacceptable risk for such literal handfuls of receptors[3], the back-loaded question would ask if we would we bother to invoke costly cleanups to protect these. The answers to the front- and back-loaded questions are (and should be) obvious. Let us return to our alternative approach to exposure assessment and the well-intended question we arrived at earlier: "How many species representatives are we concerned about for any given contaminated site?"

We should be concerned about health risk matters for ecological species at those sites where we have reliably present, the elements to satisfy the definition for a population, namely potentially or actually interbreeding individuals capable of producing viable young. A site might have two species representatives present, but closer inspection might show these to be of the same sex, and with that, no possibility of young production. Three or four species representatives might better the chances of satisfying the definition of population, but it still cannot provide the surety we are looking for. We could again have a skewed sex ratio, and it could also be that some members of this miniscule group are senescent, and thereby no longer reproductively capable. With such minimal numbers, we cannot discount the possibility that with a certain frequency, an animal or two might venture beyond the site boundary. Strictly speaking, these would not be animals that our ERAs should want to consider because they do not display the degree of site fidelity that we would prefer to have. Unfamiliar as ERA practitioners might be with a notion of screening sites for worthiness in submitting to ERAs on the basis of the presence of a threshold count of species representatives, none should object to setting such a threshold at just five individuals (Tannenbaum 2005b). Let us put this reasoning into practice.

A site gearing up to have an ERA done is located in a region where badger (*Taxidea taxus*) occurs. The contractor to assemble the ERA is considering selecting badger as a potential ecological receptor of concern. His cursory review of the literature reveals an average density for the species of five individuals in 247 acres. The site that is to have the ERA carried out though is a mere 12 acres in size. The contractor learns that there simply cannot be enough badgers present at the site to legitimize that species' inclusion in the ERA. Additionally, now stored away for future reference is the knowledge that a terrestrial site that is any smaller than 257

[3] An impossibility still to the present day because no expression exists for ecological risk, statistically, mathematically, or otherwise.

acres won't include badger as a receptor of concern. The easily acquired density information can be put to still better use. From this point forward, contractors assembling ERAs for sites smaller than 257 acres, and situated in regions where badger occur, could take care to deliberately omit any mention of badger being known to reside in the region, true as it might be. How streamlined the ERA process would be if technically unsupported requests to include certain receptors didn't have to be addressed. Over the years we have surely learned that the mere mention of a species in an RI/FS's site description section, or in an ERA's narrative pertaining to the site setting, is enough to trigger such requests.

Of course, the intent of what we might term "receptor presence-based screening", is to review the exhaustive list of animals at a site in the manner just described. Operationally then, a site species list should first be assembled. Commonly, ERAs provide lists indicating species that have been observed onsite, as well as species that are assumed to be onsite but have never actually been observed there. Accessing animal density information to facilitate the screening process poses no challenge today, given the plethora of reliable source information available in the open literature and on the internet. Density information for mammals will commonly be reported in units of animals per acre or hectare. Bird densities will often be reported as pairs per acre or hectare. One should expect to come across somewhat divergent density figures for any given species, recognizing that this variation reflects such things (among many other data influences) as the specific research technique employed when gathering the data, the level of effort expended, the habitat features of the study area, the season(s) during which the research was conducted, and the geographic locale involved. Given the variation one is likely to note in a compilation of density figures, a recommended approach is to adopt the highest reported density. Where a highest figure reflects only one sex, a singular season studied, and a US state far removed from that in which the contaminated site of interest is found, the extreme conservatism associated with utilizing the figure in screening is manifest.

Receptor presence-based screening need not be a task to be undertaken anew at the outset of each ERA, and certainly in the case of mammals. Recognizing that there is a finite number of mammalian species routinely encountered and evaluated in terrestrial ERAs, for simplification purposes a fixed table can be constructed to assist screening efforts. Table 3.1 constitutes such a list, citing species' average and maximum densities. The reader should note that average density information populating a look-up table can be derived in two ways. The ecological risk assessor may elect to comb the literature on his/her own, proceeding to then average the mean figures compiled. Alternatively, the reviewed literature may be found to report average species densities of its own. In the former case, the would-be assessor needs to decide if his/her review was extensive enough to justify the computation of an "overall" average density figure.

Table 3.1 Species density figures for use with receptor presence-based screening (mammals).

Species (common name)	Reported density (animals/acre)	
	Average	Maximum
Black-tailed jackrabbit	0.05	14
Coyote	0.001	0.009
Eastern cottontail rabbit	2.38	23.9
Long-tail weasel	0.03	2.8
Mule deer	0.04	0.2
Raccoon	0.013	0.61
Red fox	0.006	0.03
White-tailed deer	0.05	0.04

The reader will come to realize that most, if not all sites screened in the manner that has been discussed haven't a legitimate mammalian species to assess from the vantage point of sheer numbers of animals present. A similar conclusion will undoubtedly and commonly be reached for birds as well, once the earlier described essential front-loaded and back-loaded questions have been resolved for *pairs* of a given species. It is the manifestation of such realizations that affect the ERA process in several dramatic ways. Elementally, a finding that there aren't enough site receptors to matter, raises an awareness that ERAs as they conventionally take form, are unnecessary undertakings. This is remarkably good news. Where we may have initially resigned ourselves to being party to unavoidable and convoluted multi-year projects that hope to determine the necessity of site remediation, we instead stand to learn that the prerequisites for assessments aren't present in the first place. Where we allow ourselves to recognize that sites are commonly marked by an insufficiency of receptors, we have the good fortune to sidestep HQ computation and with it, all the difficulties that this mismeasure presents (Tannenbaum et al. 2003b; Tannenbaum 2005a). We are all too familiar with ERAs that produced failing HQs and that also concluded that the sites did not pose health effects because their small sizes didn't allow for sufficiently robust populations to live there. Traditionally such dichotomous reporting has left many wanting for an explanation of the supplied HQ computations. Surely it was apparent *before* the computations were done that the site's miniscule size precluded a sufficiency of receptors being present. Seemingly we are forced to say that either force-of-habit brought along the computations, or more to the point, our insistence that the conventional approach to ERA is so correct and utilitarian, that we cannot conceive of an occasion when the standard fare would not be welcome. Until an improved replacement paradigm comes about, we should be concerned with those assessments at insufficiently sized sites that fail to report their insufficient animal numbers as a consequence. Insufficiently sized sites where "area use" was taken into account have a better chance of producing HQs below unity. It is here that errors in site analysis stand to be recorded for posterity. In fact, it is not the outcomes of HQs <1 that allow for conclusions at insufficiently large sites to be designated as not posing health risks to receptors. Such designations follow from the sites not housing a sufficiency of receptors such that an HQ calculation should have taken place altogether. In this context, it is quite worthwhile to note that there is no absolute requirement to compute HQs as a/the means of adjudicating sites (regarding their potential to pose harm to receptors). Fully valid and far more utilitarian assessments for the sites discussed here, could take the form of no more than one-paragraph narratives that explain, on the basis of an insufficiency of biota, the absence of a need to submit to any structured ERA exercises.

The concept of an insufficiency of biota takes two distinct forms. With the foregoing discussion of animal density, we have discussed the first. In introducing the second form – spatial movements/home range characteristics – we find that for the receptors we observe at sites, even if there are enough of them (i.e., more than perhaps five), there is insufficient site contact to the point that health concerns are legitimized. Prior to delving into a brief analysis of home ranges, we might pithily confront our ERA lot this way: it is no one's fault that contaminated sites that stand to be ecologically assessed are relatively small, just as it is no one's fault that naturally, the species we have an interest in protecting have relatively large home ranges. The combined realties come to supply what should be pleasing news. Vis-à-vis site-specific ERA interests, the primary subject of this book (as opposed to large/watershed-scale or regional interests), there is little work to be done; a great many sites fall away. Of course, where bias intercedes and where concern over job security in the ERA field is perceived as threatened, demonstrations of receptors not being sufficiently exposed to chemicals will be ignored, and the realties spoken of here (of sites being relatively small, and home ranges being relatively large) will be overlooked.

A useful exercise paralleling the tabularization of animal density figures described earlier, would entail a one-time tabularization of the home ranges of receptors likely to be considered in conventional desktop-based ERAs. (A cautionary note to the reader is appropriate at this juncture. The US EPA's Wildlife Exposure Factors Handbook (USEPA 1993) does not furnish a sufficiently complete review of animal densities. By way of example, excluding the three small rodents and one mammalian insectivore that are reviewed (because site remedial decisions are not based on such, or should not be), only a scant seven other mammals are cursorily reviewed.) Such an effort will undoubtedly have the effect of crystallizing for the user, the awareness that in a great many cases, receptors do not sufficiently contact sites such that they would develop critical toxicological effects. Here the reader would do well to ponder, *a priori*, how much of a receptor's time spent interacting with a site's contaminated media is needed to elicit a toxicological effect, or to justify a species' selection as a receptor of concern. It is recognized that making such inquiry parallels the earlier posed question of how many species representatives need be present to either legitimize conducting an ERA or taking remedial action. Both are questions that are never asked, and are questions that ERA practitioners would undoubtedly prefer not to have to answer. We may speculate here on a critical point to be elaborated on later (principally in Chapter 7). It is neither the absence of guidance indicating the minimum degree of site contact/site presence needed to trigger an ERA, nor the inability of ERA practitioners to come forward with an answer, that would account for reticence when questions of this order arise. We may suggest that the aversion of ERA practitioners to fielding such questions reflects their not wanting to hear, or be convinced, that conventional ERA treatments are unnecessary. Electing here nevertheless to address head-on, this new "required degree-of-site contact-to-elicit-a-toxicological-effect-of-concern" question, can serve to well educate us.

Table 3.2 reviews home range information for a number of mammals that are commonly included in ERAs. As with the animal densities, and as we should again expect, the literature reports a fairly wide range of estimates, reflecting the efficacy and scope of the studies (e.g., study duration, number of animals involved), geographical variation, the season(s) considered, etc. In addition to the estimated values from field observation (expressed as both a range and an average), the table provides estimated allometric home ranges (Harestad and

Table 3.2 Home range figures for use with receptor presence-based screening (mammals).

Species (common name)	Home range values (acres)		
	Reported range	Reported average	Calculated (allometric) minimum
Black-tailed jackrabbit	40–49.9	7.41	49.9
Coyote	3,532–16,796	2750	16,796
Eastern cottontail rabbit	3–20	10 or less	20
Kit fox	84–1,330	30–40	1,330
Long-tail weasel	12.35–298.9	17–37 (m); 3–10 (f)	298.9
Mink	640–2,470	1900	2,470
Mule deer	90–600	–	600
Opossum	10–50	30	50
Raccoon	39–2,560	1976	2,560
Red fox	123.5–7,410	960	7,410
White-tailed deer	319–1,280	640	1,280

Bunnell 1979), i.e., modeled values based on correlations of body weight and spatial movement.[4] As a conservative gesture regarding these latter figures, we might elect to use the smaller of either (a) the lower end of the range of reported (observed) home ranges, or (b) the calculated home range. Consideration of the red fox (*Vulpes vulpes*), a frequently assessed mammal in terrestrial ERAs, will demonstrate the species' non-relevance to site-specific assessments (had we not already recognized this with the assistance of the earlier animal density analysis). The reader is reminded that Table 3.1 indicated that a site would have to be in excess of 167 acres, and more likely 833 acres, to have reliably present the requisite five species representatives that warrant a (species) evaluation. Even if a red fox has a home range of 123.5 acres (i.e., the lowest literature-reported home range; CH2M Hill 2001), but the contaminated site under scrutiny is only 10 or 20 acres, we will need to come to grips with such an areal mismatch. How a fox precisely allocates its time in the utilization of its space (here, 123.5 acres) is unknown. Nevertheless we are not so naive to think that a fox would utilize every inch of its home range equitably. Nor are we so naive to expect a fox to spend 80% or perhaps 90% of its time in a 10- or 20-acre area, with the remaining 10% or 20% of its time occupying the remaining 103.5 or 113.5 acres of its home range. Even with fox, a denning mammal, it is unrealistic to expect to see such a skewed partitioning of time. It goes without saying that ERAs that, for what are said to be conservative considerations, assess red fox and other wide-ranging mammals and birds as spending all of their time within a 5- or 10-acre space, are worthless because they do not describe real-world exposure scenarios. Such efforts bring no conservatism to the ERA process, and rather serve only to overtly telegraph regulator bias. With regulators so badly wanting to show that contaminated sites pose health effects to biological species found there, and wanting to have just cause for invoking remedial actions, unrealistic exposure scenarios are wont to materialize in conventional ERAs.

Unlike the situation with the red fox, we are fortunate to have at our disposal, a reliable and working knowledge of the precise space (and time) utilization of another commonly assessed terrestrial mammal, the white-tailed deer (*Odocoileius virginianus*). Global positioning system (GPS) technology harnessed in an intelligent way to specifically review the reasonableness in identifying this species as a receptor of concern (and admittedly to confirm what seemingly should be intuitively obvious[5]), was conducted (Tannenbaum et al. 2013). GPS

[4] For conservatism, lowest literature-reported body weights were used in generating the values.

[5] ERA practitioners who professionally may not be wildlife biologists, field ecologists, or mammalogists, are nevertheless educated sufficiently to know that white-tailed deer have commonly reported home ranges of 600 acres or more. Common sense should dictate that the white-tailed deer, due to its inherent biology, will not spend adequate time at Superfund-type sites of say 5, 10, or 20 acres (which describe the great majority of contaminated sites) to develop chemically mediated health impacts.

collars were placed on deer who were allowed to roam freely for approximately a full calendar year. Does as opposed to bucks were collared, with the former having smaller home ranges (DeYoung and Miller 2011). The intent was to try at all costs to identify instances where a deer might be found to be spatially relevant in a context of the size of contaminated sites. The collars were programmed to record latitude/longitude locations approximately 35 times/day (i.e., once every 40 minutes). The average number of locations recorded over the tracking period was in excess of 7200, with some does having in excess of 10,000 locations. Importantly then, the patterns of the plotted locations (Fig. 3.1) are incontestably descriptive of the does' spatial movements; the sheer magnitude of recorded locations for any given deer far outweigh any uncertainty associated with periods throughout the tracking study when locations were not recorded. Moreover, the magnitude of recorded locations validate the study's central underlying assumption, that greater densities of plotted locations are reflective of greater percentages of time

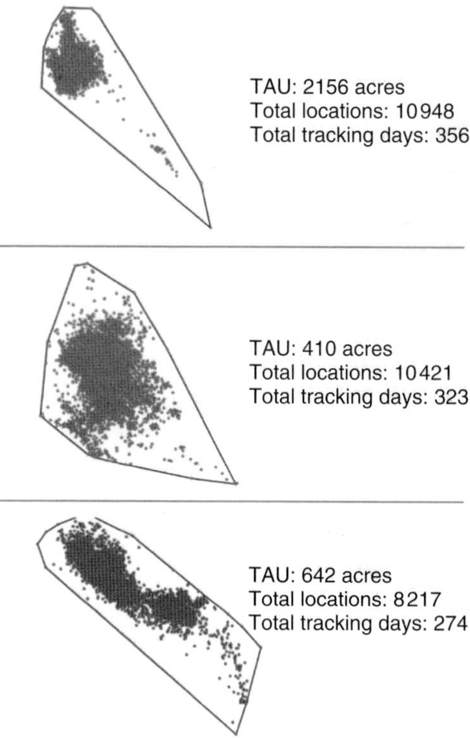

Fig. 3.1 Spatial movement plots of freely roaming GPS-collared female deer (*Odocoileius virginianus*). Denser areas of dots (precise latitude-longitude locations) are areas of greater usage/site presence. (Adapted from Tannenbaum et al. 2013. Reproduced with permission of John Wiley & Sons.)

spent by a doe in a particular area. Where the maps of the plotted locations are gridded into uniform cells of 1, 5, 10, and 20 acres to simulate run-of-the-mill contaminated terrestrial Superfund-type sites, we have the opportunity to identify "most occupied/preferred areas", and to learn of the percentage of a deer's time spent utilizing such spaces. Figure 3.2 depicts the total areas utilized (TAU) of several deer tracked for upwards of a year, where the TAU was gridded into parcels of 20 acres. In this approach, the specific method of home range estimation is not vital; the emphasis rather, is on considering the entirety of the terrain each deer covered (known from the recorded array of latitude/longitude locations) for a deeper analysis of the utilization of certain subdivisions of that greater land parcel. Figure 3.1 typifies the space utilization of white-tailed deer that have now been observed in two (geographically distanced) regions of the eastern US. Relevant summary statistics presented below (Table 3.3) for more than a dozen tracked does (of the most recent application of the above-described study design; Tannenbaum et al. 2013) go a long way towards demonstrating that white-tailed deer will consistently be inappropriate ecological receptors of concern.

There is great potential for both the summary statistics below and the pictorial information of Fig. 3.1 and Fig. 3.2 to be incorrectly applied. It is true that each doe utilized one cell (or a very few cells) more than any of the others, and independent of the overall terrain covered having been gridded into uniform subdivisions that approximate the sizes of Superfund-type sites. This finding cannot be reviewed in a vacuum, however. For the overly aggressive risk assessor (perhaps representing a regulatory agency) who is insistent that deer be an evaluated receptor in a given ERA, we can anticipate an argument put forth: "See? There *is* a cell that is occupied more than any other. And not only that – that most occupied parcel could well turn out to be a contaminated site in need of review".

The anticipated argument ignores critical information. No one will dispute that each deer had a preferred area. At the same time, we cannot fail to note that the greatest degree of occupancy for any of the land subdivisions (i.e., even the most reasonably sized ones of 20 acres) is only 32%, meaning that 68% of the time, a deer is absent from this "preferred" cell. Surely ERA's goal is not to select receptors to evaluate that are absent from sites more than they are present at them, and by an overwhelming margin no less. Coupling this minority of time spent at an area that could be a contaminated site, with the likelihood – of just 2.3% – of there being a most utilized site overall, shows us dramatically that the white-tailed deer will probably never be a reasonable receptor to consider. And there's more. With this exercise we have the luxury of reviewing real-time spatial movement data for a terrestrial receptor of renown; we are able to know daily, over a year's time, where deer actually migrate, etc. How tragic it is then, in the aftermath of a presentation of such data at a professional society conference focused on the potential chemical exposures of animals, to receive the audience comment:

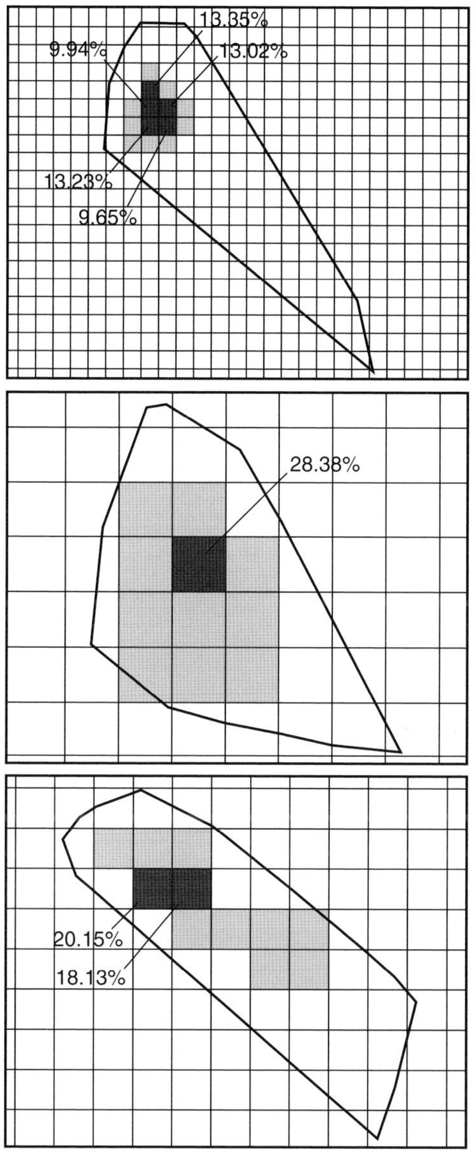

Fig. 3.2 The preferred/most utilized 20-acre (8.1 ha) cells of three profiled deer from Fig. 3.1 are indicated along with the percentage of each deer's total recorded locations that fell in each cell. The frequency of there being a most preferred cell within a deer's total area utilized (TAU) is approximately 2.3%. If there should be a sizeable (20-acre) area of contamination within a deer's TAU, a deer doesn't spend enough time there to matter. (Adapted from Tannenbaum et al. 2013. Reproduced with permission of John Wiley & Sons.)

Table 3.3 Demonstrated spatial irrelevance of white-tailed deer at contaminated sites (following the analysis associated with Fig. 3.2). (Adapted from Tannenbaum et al. 2013. Reproduced with permission of John Wiley & Sons.)

Computed statistic	Value (%)
Probability of there being a highest occupancy 1-acre cell	0.004
Greatest degree of utilization (occupancy) of a 1-acre cell	Maximum: 4.98; Mean: 3.4
Probability of there being a highest occupancy 5-acre cell	0.02
Greatest degree of utilization (occupancy) of 5-acre cells	Maximum: 17.2; Mean: 10.0
Probability of there being a highest occupancy 10-acre cell	0.02
Greatest degree of utilization (occupancy) of 10-acre cells	Maximum: 26.6; Mean: 16.3
Probability of there being a highest occupancy 20-acre cell	0.023
Greatest degree of utilization (occupancy) of 20-acre cells	Maximum: 32.2; Mean: 25.5

"Have you thought about *modeling* the deers' spatial movements?" Can modeling possibly hold a candle to having the good fortune of knowing how an animal *actually* behaves in the wild? Remarks such as the one mentioned here need to be heard in the correct vein; effectively, the request to try our hand at modeling is communicating an unwillingness to hear that we can actually *know* that certain species are either spatially irrelevant, insufficiently exposed, or both.

We should not be so naive to think that it is only deer, as a wider-ranging receptor, that may not be spatially relevant for the sites with which we commonly deal. If we could collar red fox, mink, long-tail weasel, or raccoon in a similar way, we should expect to compute similar statistics. In conjunction with animal density information that is easily assembled, we are faced with the reality that chemicals are not bioavailable in any meaningful way simply because animals do not encounter them sufficiently in the first place.

The reader is reminded of the commonplace difficulty that birds present in ERA, specifically when migratory species are proposed and selected as receptors of concern. The familiar case is one where an overwintering species is absent from the site for approximately 4 months of the year. Although the bird is site-present two-thirds of the time, what does this say about its susceptibility to chemical attack, our chances to observe signs of reduced fitness, or to establish that the site is at fault for any reduced fitness we might observe? Granted, we are speaking here of a *majority* of an animal's time (still) being spent onsite, but what occurs during the overwintering months is unknown to us. Conceivably overwintering birds might be contacting contaminated media once again. For that unfortunately most rare of cases, where site-specific ERA efforts advance to the point of direct field observation (for such things as general appearance, nest counts, and clutch size), we run the risk of wrongly ascribing what may be termed as health impacts, to contaminant exposures that stem from locations far removed from the site. Non-year-round species are less than desirable for a second reason. Assuming their remote overwintering locations are relatively pristine ones, the birds are then afforded an uninterrupted 4-month (or greater) stretch to clear their bodies of potentially harmful contaminants they might have assimilated while occupying the contaminated site. Reasonably too, it could be that site birds are seasonally sickened, but manage to repair their health condition over the months spent away from the site.

The purpose of this discussion has been to raise the awareness that birds, like mammals, will often be contraindicated as receptors of concern. Due to their migratory habits and their generally low species-specific densities, we are faced again with the reality of their not being sufficiently exposed to contaminated sites to matter in a risk assessment context. One way or the other, individual animals are not contacting contaminated sites to a point such that we could anticipate their succumbing to the chemical stressors that are presented to them. Suggesting that we, from this point forward, divest with including birds in conventional ERAs is a bold proposition but one that has scientific merit. Alas, it is a proposition that would never be given due consideration by the powers-that-be.

Earlier in this chapter, the would-be concern over toxic exposure at contaminated site settings was recast a number of times. The question of how much of a chemical in a medium is free to be taken up by a receptor was reformulated to ask how large a site we might be considering. This was done to recognize that there might not be a sufficient supply of animals at the site to create much of a stir. This thinking led to two other questions, one asking about the (minimum) number of animals needed at a site to truly garner our attention, and the other seeking out the actual number of animals utilizing a given site. For this last query we can always handily supply an answer, for there is no shortage of published information on animal densities. We would do well to take a step back from the bioavailability/exposure assessment element of ERAs, and to allow the imperative question to morph still another time. In the light of the relatively large parcels needed to house enough receptors to make a go of it, and/or to ensure that receptors spend enough of their time contacting contaminated media, we need to ask "What contaminant release event on our planet would have to occur in order to produce a sufficiently large contaminated area to make ERA meaningful?" Let us return to our earlier red fox example, in light of this species having become a veritable mammalian fixture of terrestrial ERAs. With a five-animal area of 833 acres based on *average* reported densities (as reported in Table 3.1), how could a contaminated area this large come about? The answer is that for all intents and purposes, it could not. There are, of course, mine tailings sites that extend for tens of miles or more, but these do not in any way typify the sites that ERA deals with in the main. Even so, we must acknowledge that where such areally-pronounced sites occur, fox may not present. There are too, extensive agricultural parcels that may have been irrigated with contaminated water for long periods. These may supply the acreage to make the five-animal area a possibility, but corn and alfalfa fields and the like do not describe the habitat with which fox are associated. To make the five-animal area a viable consideration, and perhaps to soften the blow for the reader who is unwilling to hear that fox may never be a site-relevant receptor in a context of Superfund-type sites, let us consider the fox's five-animal area based on *maximum* reported densities. What situation would give rise to 167 acres of land being contaminated throughout? The engaged reader might appreciate first putting an area of 167 acres into perspective, for the relevant discussion is one of scale.

Although not uniform in their dimensions, all major league baseball fields have roughly the same layout. The span from home plate to the extremity of center field is approximately 375 feet. The extremities of the right and left field foul lines are approximately 320 feet from home plate. From home plate to the backstop is generally about 60 feet in distance. Considering these measurements and including foul territory, the total playing surface of a major league baseball field is somewhat just shy of 6 acres. If we can imagine a contiguous area comprised of 28 such baseball fields we will then have a representation of the required contaminated land surface to support a red fox assessment (something the book's back panel was intended to illustrate). It would seem that a contaminant release event to chemically disaffect an area the size of 28 major league baseball fields would have to be cataclysmic. The reader would do well to consider the acreage of what can be termed "the World Trade Center site" in New York City's lower Manhattan district. The complex's seven buildings that either toppled on September 11, 2001 (prominently among them the majestic 110-story twin towers) or that had to be subsequently demolished due to their instability just days and weeks following the attack, occupied 16 acres of real estate. If the highly coordinated September 11 terrorist attack could damage no more than one-tenth of the area needed to propel us into a fox-based assessment, the prospects for less dramatic contaminant release events to result in sufficiently sized contaminated areas are probably not there. Before examining this phenomenon, let us set the record straight regarding September 11 *aerial* releases, since well-documented dust cloud deposition extended for tens of miles. Although the contamination footprint resulting from the deposition was surely large enough to make relevant any species' assessment, we must recall that the aerial releases of industrial sites that become Superfund fare, never travel such distances. A case in point is a former secondary lead smelter in New Jersey that operated for a decade and that gave rise to a 46-acre National Priorities List (NPL) site. Curiously the disaffected acreage at that site reflected more than just the area "dosed" by the aerial releases; it also included the lead contributions of dumped car batteries, process wastes, and still other waste generation forms. Importantly, if the entirety of the 46-acres had the smelting operation as its sole source of lead contamination, there would nevertheless be an insufficiently large area for ERA purposes.

The well-characterized NPL universe can confirm for us that the routine (and not cataclysmic) contaminant releases we know of, do not give rise to sites that are large enough for ERAs to consider. Figure 3.3 examines the single largest waste-generating "industry" that has influenced the NPL universe, namely manufacturing (US EPA 1991). With contributions to the universe of 52%, it handily surpasses mining, municipal and industrial landfills, the military, the Department of Energy, and recycling; its next nearest rival is municipal landfills, with a 23% contribution (US EPA 1991). The 13 manufacturing categories profiled in Fig. 3.3 indicate the relative contributions of these to NPL sites, and also allow us to consider how physically large the facilities churning out a wide array of products might be. It should be evident that individual establishments for any of the categories do not occur on properties of 50 or 100 acres. If they did so, it would not be the entirety of the property that became contaminated should unintended chemical releases

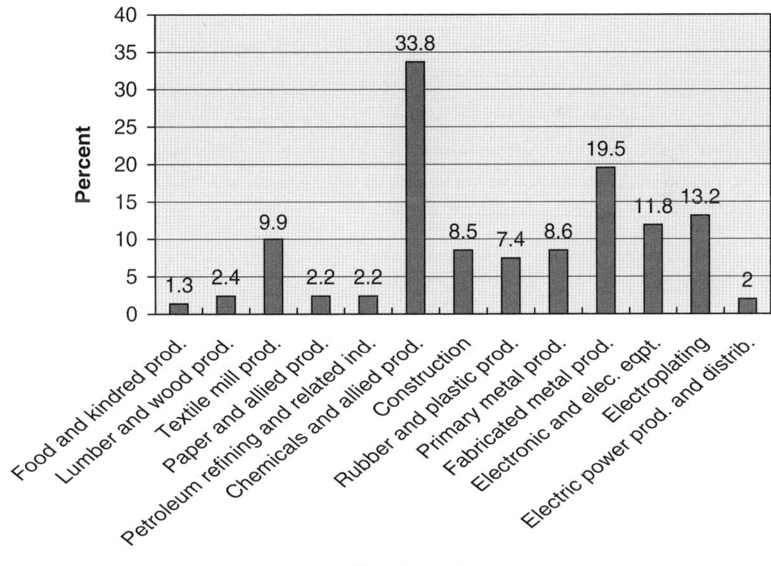

Fig. 3.3 Manufacturing, the single largest waste-generating industry influencing the NPL universe. In addition to quality habitat to nurture ecological receptors (that could form the bases of cleanup actions) being lacking, the acreage associated with plants for the various industries is unlikely to be sufficiently large to legitimize site-specific ERA interests. (United States Environmental Protection Agency (1991).)

(e.g., spillages) occur. The extent of viable habitat to support ERA needs is critical to the discussion of manufacturing plants. Although the plants might abut usable habitat, the plants themselves would encompass buildings, roads, equipment, and asphalted surfaces. Additionally, the plants would invariably have considerable vehicular traffic and noise generation associated with their operation. All of these features would make the plants unlikely to lure to them the species that regularly concern us in ERA.

We will borrow the red fox one last time for a hypothetical and consolidated consideration of scale within the framework of exposure assessment. A group of 25 well-educated and highly experienced field ecologists, wildlife biologists, and mammalogists is assembled for the task of identifying a 100 miles^2 area in the United States, distanced at least 50 miles from any known site of soil and water contamination, where red fox should, with the best of assurances, be the most numerous. Over a period of a year, the group is encouraged to seek out any and all information that can influence their judgment, such as fox census records, predator-prey cycling data, and hunting statistics. State-of-the art mapping is made available to the group for all those factors that can influence red fox population size, including such things as habitat type and quality, geographical terrain, human interference (e.g., highways, land development), past rainfall records, and overall weather conditions. After the group identifies the 100 miles^2 area, 25 randomly selected 10-acre parcels are identified, and at each one, 3000 surveillance cameras are mounted from tree-tops to monitor fox activity for an entire year. The cameras are angled in such a way that the entirety of each 10-acre parcel will be scanned. So as not to bias fox movements, the camouflaged cameras are treated to remove all traces of human scent that might linger on them from their having been handled at the time of deployment. After the year of surveillance photography is concluded, the films are reviewed. The reader is, at this point, asked to realistically anticipate how many different foxes were observed at each of the 25 locations, even if the totality of any one fox's time spent within the boundaries of a 10-acre cell (over the course of the monitoring year) was no more than 1 minute. A common sense approach to answering the question of how many different fox appeared at any of the parcels, would require nothing more than recalling that the red fox has a naturally sparse distribution in nature, and then simply extracting a maximum animal density figure from the open literature. In this specific case, with a maximum density of 0.03 animals/acre, not even one red fox would be expected to be viewable in a frame-by-frame review of the year-long footage taken at any of the randomly selected 10-acre sites. The same would be true if the surveillance work had taken place at randomly selected 20-acre cells, and still bigger ones. A failure to observe in the wild, animals or their tell-tale signs (tracks, scat), does not mean that contamination is precluding their appearance or their ability to establish a community. It means that a species' spatial habits were not correctly taken into account. One should not expect to see a wide-ranging receptor at a typical contaminated site working its way through the remedial process, just as one should not expect to see a wide-ranging receptor at a comparably sized pristine parcel. For the ecological risk assessor, the task remains to become fluent with the concept that the exposure component of the essential risk equation, risk $= f(exposure) + (toxicity)$, may drop out. Further, when faced with the prospect of not having a receptor to evaluate (owing to spatial considerations), it is poor practice to adopt a "but the show must go on" approach. Deliberately selecting what are knowingly substandard receptors (as in those that are not spatially relevant to the site) is deeply damaging to ERA. To the contrary, there is no shame or embarrassment in acknowledging that a site doesn't have receptors in need of assessment. It is worthwhile reviewing a poor choice example from a Department of Defense ERA from the early 2000s. The defense site, part of an atoll complex in the South Pacific, is an island that has been enlarged over a period of years through construction activities. The only non-human mammal found on the island is the house mouse, with this species bearing the curious distinction of being the focal point of an aggressive pest eradication program. Since the mouse in this case fails to satisfy the essential criterion of an assessment endpoint, i.e., a valued resource that is to be protected (US EPA 1992a), and since no other non-human mammal existed at the island (as is still true presently), a mammalian assessment was not sanctioned. Regrettably, the mouse was selected as a receptor (of concern?) because the regulatory powers-that-be let get in the way, their non-familiarity with the notion of knowingly dispensing with the inclusion of a mammalian receptor when a reasonable case to consider is absent.

The well-characterized NPL universe provides revealing statistics regarding this site size discussion. Table 3.4, adapted from www.epa.gov/superfund/news/benefits/ch3.pdf, embellishes the earlier mentioned statistic of 60% of NPL sites being 20 acres in size or less. The reader will note that some 70% of the sites are

Table 3.4 Categorization of Superfund NPL sites by size. (United States Environmental Protection Agency (1991).)

NPL site sizes (acres)	Percentage of total NPL universe
<1	6.4
1–10	24.5
10–100	40.5
100–1,000	19.3
1,000–10,000	10.9
10,000–100,000	3.2
>100,000	0.6

100 acres or less, and that some 90% of the universe is comprised of sites that are under 1000 acres. Reported site sizes of 1000 acres and greater can be misleading in at least three ways: they can suggest that continuously contaminated properties of huge proportion exist, that singular ERAs are conducted for sites that extend to the multiple thousands of acres, and that terrestrial acreage is that being spoken of. None of these notions are true. Let us consider that if there *would be* a continuously contaminated 2000 acre land parcel, the soil sampling effort in support of a conventional HQ-based ERA would be prohibitively expensive. Even if the sampling rate stood at just 1 sample/acre, a rate that would *not* constitute a reliable areal characterization of the "site", 2000 samples would be needed (apart from required background and quality control samples). If in an effort to reduce this unmanageable sample number, it was decided to composite samples (e.g., have one sample aliquot taken from the combined soil samples of four contiguous acres of land, thereby compressing the sample universe to 500), the uncertainty associated with chemical detections would be even more excessive, rendering the sampling effort rather pointless. In summary, sites that (in theory) extend to the hundreds and thousands of acres cannot be sampled for risk assessment purposes in any practical sense. This is of no actual consequence, however, because realistically the prospect of such enormous continuously contaminated sites does not exist, as was elaborated on here.

Regarding the Table 3.3, nearly all of the sites that either approach 10,000 acres or exceed this size are federal facilities. The figures then are referring to the overall acreages of Department of Defense installations, and not to the sizes of parcels that will have (individual) ERAs done. Here it is important to note that installation-wide ERAs, *in the literal sense*, do not ever proceed. Instead, individual areas of concern on an installation, with distinct current and/or historical usages, are those that submit to ERAs. (Only when all of the individual-area ERAs have been completed, can it be said that an "installation-wide" ERA has been conducted.) By way of example, the Army's Aberdeen Proving Ground (APG) in Maryland has a combined total acreage for its two land masses of 72,500 acres. Even for APG's smaller portion (Edgewood Area), itself an NPL site, there has not been an ERA carried out to address its 17,000 acres all at once. Focused ERAs for Edgewood Area have addressed contamination at specific testing ranges and the like, and these have had highly manageable acreages (i.e., they have only occasionally exceeded 100 acres). It is also important to note that it is a commonplace occurrence that vast Department of Defense installations have their terrain, with any potentially affordable ecological habitat that may be included, interrupted due to mission work. In the case of APG, over 2000 buildings dot the installation, and cantonment areas, golf courses, and other paved surfaces and obstructions significantly subtract from the available area that can rightfully submit to ERAs.

The third way in which (large) contaminated site size statistics can mislead occurs when there is an under-reporting of information. Often the acreages provided in site summaries and in hazardous waste site databases reflect extending and widening plumes of contamination in one or more aquifers. While the soil atop at least a portion of lengthy groundwater plumes may undeniably be contaminated, it is not the terrestrial element of the site that is the big contributor to the chronicled spatial statistics. This pattern of contamination is of great consequence in ERA. Ecological receptors do not contact groundwater directly. It is only in that case where groundwater discharges to surface water that contaminants in the former can present themselves, usually in highly diluted fashion, to fish and other aquatic ecological receptors.

In the foregoing treatment of scale we have learned that truly areally-large contaminated sites rarely form, even where cataclysmic events occur. To complete our spatial analysis and to secure a heightened appreciation for the exposure term in ERA often being inconsequential, a review of how typical Superfund sites (both NPL and non-NPL) form, or more correctly, have been formed in the past[6] is provided here.

- Site 1, located in a wooded area of undeveloped land in the New Jersey Pine Barrens region, became an NPL site pursuant to a singular chemical release event that occurred in the late 1970s. An individual possessing approximately 200 containers (mostly 55-gallon drums) of solvents, paint, and paint sludges, convinced the property owner to dig a pit on his 1-acre property so that the containers could be dumped there. Following the dumping incident and the later deterioration and leakage of the containers, a 2000 square foot portion of the landowner's property (i.e., an area measuring approximately 45 feet by 45 feet) displayed visibly stained soil. The dumping incident, not surprisingly, also gave rise to a complex meandering groundwater plume of volatile organic compounds. This contamination necessitated extensive groundwater monitoring and an equally extensive treatment train to return groundwater in the area to a drinking standard safe for humans. Remedial action to address the terrestrial component of the site involved the removal of 40 yards of drummed material, 8 truckloads of excavated soil, and approximately 3000 gallons of liquid material.
- Also located in the northeastern US, Site 2, is located on a 34-acre property zoned for light industry, and with a habitat that includes significant drainages to a swamp ecosystem. From 1961 to 1975 a tank-washing operation

[6] For all intents and purposes, the exhaustive universe of severely contaminated sites in the US, numbering in the 1400s, has largely been identified for some time. Opportunities for new Superfund sites to form in the present day are minimal, owing to the increased environmental awareness that uncontrolled chemical releases can harm human and ecological resources, and the existence too of strict legislation that discourages wanton environmental offences such as deliberate chemical release.

allowed thousands of gallons of wastewater generated from the cleaning of tanker-hauler vehicles to be deposited in a series of seven unlined lagoons. The lagoons discharged to a series of swamps and creeks fringing the property, and subsequently influenced the groundwater. This in turn, jeopardized the health of the nearby township population of some 3000, who had been relying on well water for drinking, bathing, and washing clothes. Aside from an extensive groundwater evaluation, and then the construction and implementation of a groundwater extraction and treatment system (similar to that required for Site 1), the selected site remedy involved the excavation of approximately 7 acres of contaminated sediments and soils in a wetland area.

To the extent that the typical Superfund site can be characterized, the two site descriptions capture the salient features of such an entity. First, the contamination is decades-old. Discussed in the earlier chapters, this could likely mean that ERA is altogether contraindicated. For the specific discussion here, the spatial pattern of contamination is key. Owing to the natural force of gravity, and aided by numerous site-specific factors that include depth to groundwater, soil type, annual rainfall, and of course, the types and forms of the released chemicals themselves, there will almost always be a groundwater component to be addressed. In stark contrast, the spread of contamination in the horizontal plane is characteristically limited. Although there do exist occasional contaminated terrestrial parcels that encompass several hundred acres or more, such as the previously mentioned mine tailings sites, these are very much the exception; the overwhelming percentage of terrestrial sites that submit to ERAs are vastly smaller than this. Let us conclude this analysis by submitting to greater scrutiny the earlier mentioned statistic that nearly 60% of NPL sites are 20 acres in size or smaller (US EPA, 1989a). In addition to the 20-acre figure referring to a groundwater site component, the acreages we read about may refer to the property-line boundaries of the overall owned parcels as opposed to the actual contaminated "sites" of concern within that parcel. Site 2, described earlier, demonstrates this last point. In total, the industry's property covers an area of 34 acres. It was just 7 acres of land surface however, that either received the many gallons of chemical rinsate by direct addition to the ground, or that became contaminated through indirect means such as overland flow and seepage events.

With regard to terrestrial sites, the foregoing analysis has neglected to consider the exposures of small rodents. This was done by deliberate choice. Invariably small rodents will always be spatially relevant at sites. That is, their densities can almost always guarantee that there will be present, at even the smallest of sites, a sufficiency of species representatives to work with. Due to their miniscule home ranges and non-migratory behavior, small rodents have rather maximal site exposures, and only very limited opportunities to remove themselves from contaminated properties they might occupy. Small rodents though, are not profiled in this chapter because of an as yet unwritten but sorely needed rule in ERA: "Small rodents do not conjure up in the psyche of the ecological risk assessor or any stakeholder in the site remedial process, the sense of an animal in need of protection, or one for whom a cleanup could alone proceed. Small rodents and related species are not to be selected as receptors of concern." Although it might prove difficult to exact a confession of this order from a regulator, this must be what he/she believes deep down. The reasonably minded individual will understand that for a site of only 1 or 2 acres, the ecological receptor list could include nothing more than a small rodent species or two. The reasonably minded individual further understands that a finding that site rodents (at such small sites) may be "at risk" (presumably from a HQ calculation) does not inform in the least on health effects that may accrue to higher food chain species that might access the site on an infrequent basis. It is curious (if not disturbing) to so frequently encounter regulators who seem to forget that HQs are species-specific. A mouse HQ greater than one means that by estimation, the mouse is consuming more of a chemical than is safe. The mouse HQ however, reveals nothing about the welfare of higher predatory species (e.g., an owl) that might consume the mouse. To suggest as the reason to move for a site cleanup, that mice at the site are likely to die off because of their high chemical exposures, thereby imposing a dietary hardship (the absence of suitable food) on mice predators, is nothing more than conjectural matter. Rather than constructing a wishful "the sky is falling" site scenario, the regulator should be answerable to a brief set of pointed questions: What is the home range of the predatory species of concern, and what percentage of its annual quota of consumed rodents is the 1–2 acre site likely to be supplying? Has any rodent trapping been conducted at the site, and if not, why not? If rodents should be in adequate supply, is it not clear that site rodents are withstanding their site chemical exposures?

The foregoing analysis has all but neglected the chemical exposures of aquatic species. This too has been deliberate, and it follows from the unfortunate inattention given to the spatial dynamics of this animal grouping in day-to-day ERA efforts. Aquatic receptors at contaminated sites, principally fish, should be recognized substantially more than they are, specifically because of their rather unique exposure setting, namely being ceaselessly surrounded by and bathed in their supporting environmental medium. The list of terrestrial organisms that share a similar existence runs only to earthworms and perhaps burrowing moles. (Earthworms, although supplanting the diet of many trophically higher forms, are relegated to a status similar to that of the small rodents just discussed; sites are not remediated in order to afford protection to them. Moles are simply not species of interest in terrestrial site work presumably because we don't know enough about them, this a function of their veiled existence.) The standard approach to aquatic receptor assessment conspicuously sidesteps exposure assessment concerns. [While studies of the (conventional) bioavailability of contaminants in water and sediments may routinely occur, these are modeling affairs that can at best suggest that contaminant uptake is occurring. Aside from our inabilities to relate assumed tissue concentrations of contaminants to health effects, the studies overlook our need to know how many receptors (of legitimate concern) may stand to be chemically exposed.] Instead, the standard approach rather routinely proceeds to the direct comparison of contaminant concentrations in the water column and sediment, to tabular concentrations that are held to be either safe or to impose health effects. Little effort is expended prior to such comparisons, as in conducting a formal exposure assessment step – one that would report on spatial relevance, as in indicating if enough receptors (e.g., fish) are present to warrant concern. The situation needn't be this way, and enhanced efforts to characterize exposure could inform us considerably about the aquatic site dynamic.

Exposure assessment in HHRA identifies the specific site receptors that interact with the chemically affected media, the extent to which contact is made, and the receptors' specific operative chemical uptake routes. The situation is (or should be) isomorphic for terrestrial ecological receptors (hence this chapter's earlier generous attention to anticipated numbers of site mammals and birds, and the degree of their site contact based on spatial movements, etc.), and also for aquatic receptors. We would be well on our way to enhanced aquatics exposure assessment if simplistic census comparisons were conducted. In addition to identifying the specific fish species living in a contaminated stream, pond, lake, or river reach, we should know approximately how many fish of each type there are, and we should know the same for a comparable nearby waterbody unaffected by the site contaminants. If the numbers were about the same, it would (or should) be clear, at this minimal level of analysis, that there was no need for additional assessment. Finding that the site and the reference location support similar numbers of fish tells us that the site is not limiting to them, or more correctly, that the site is incapable of impacting fish numbers. There are two critical concepts here. First, if the site hasn't acted to reduce fish numbers after a decade or more, it's not ever going to. Second, if the site should be limiting to fish, we have no credible way to explain how other fish might be arriving to the contaminated waterbody to supplant a supposed reduced site population. We know that fish do not fly, and we won't believe that fish are being covertly air-lifted to the contaminated waterbody by some radical extremist faction (similar to what was said about earthworms and field rodents in Chapter 1).

The beauty of first characterizing exposure through comparisons like that described here is in helping us to avoid conducting other unnecessary efforts with all their potential to lead us astray. If the animal (here, fish) numbers are there, we have no need to see if they are living beyond their means, i.e., that they are being exposed (so some might think) to chemical concentrations that are in some way toxic to them. If a site displays a reference location-comparable population size, yet is said to be experiencing unhealthful chemical exposures as per the outcome of a toxicological review, we must believe that the receptors are fine. If a bleak toxicological review points at reproduction and not survivability as being impaired, how then are the balanced fish numbers of site and reference location to be explained?

The previous text has endeavored to reinforce the point that so long as a contaminated site furnishes the animal numbers it should, we have sufficient feedback to know that site ecological receptors are adequately protected. This is a concept for the ERA practitioner to secure. The formal ERA process currently in place (just as with the HHRA process) is not a linear one in the sense that the Exposure Assessment and Toxicity Assessment steps are to be conducted in a fixed sequential order (US EPA 1989b, 1997, 1998). It would appear that there is a distinct advantage though, to deliberately conducting the Exposure Assessment step first, for doing so can obviate the need for Toxicity Assessment and what are perceived to be other essential ERA tasks. While the ERA practitioner works at his/her "assignment" (i.e., to learn to be contented with the information supplied only by an Exposure Assessment effort), the "assignment" for those who have crafted the currently reigning ERA scheme is to become comfortable with the notion that there needn't necessarily be a Toxicity Assessment step at all. Consideration of an extreme and esoteric example will supply context to the respective assignments. Suppose site-specific field work reveals that in response to a chemical stressor, a species (or to make the case more demonstrable, several of them) has a considerably greater neonatal mortality rate than that occurring in the background, i.e., in the nearby area beyond any influence of the site's contamination. The site-specific studies reveal also that counteracting the neonatal losses, litter/clutch size for the past 10 years has been considerably augmented.[7] Although we would probably prefer to have not happened upon a compensatory mechanism at play to correct for a demonstrated site-posed reproductive effect, we should not lose sight of the fact that the site has every bit as many species representatives as it should. For this site arrangement, there would be no need to work at unraveling any stressor–response mechanisms associated with the boosted young production that was observed. Although sufficient animal numbers are being supplied through a reconfigured design, it is (or should be) enough to know that the site has its full complement of animals. We are reminded that it is ERA's charge only to see to it that animals are protected. At our fictitious site, they are. Although the site has the nuance of its young production being

[7] The situation is not really so far-fetched. Documented cases abound where birds laid additional eggs (larger clutches or extra clutches) in response to deliberate experimental intervention where nest eggs were selectively removed. The text's 'extreme and esoteric' example *is* 'far-fetched' because elaborate and highly detailed, multi-year ecological investigations rarely if ever proceed for Superfund-type sites.

atypical (perhaps), it is not ERA's charge to see to it that contaminated sites function exactly as do non-contaminated properties.

Admittedly, censusing ecological populations can prove to be costly and time-consuming. Then again, we have learned that for mammals and birds (the only two groups of terrestrial receptors ever given consideration in ERA), censusing efforts would never be needed, for we have excellent animal density statistics to access. The statistics can be expected to commonly inform that sites of interest do not house a sufficiency of animals to concern us in a great way. On the aquatic side then, we must rhetorically ask how hard is it to (approximately) ascertain how many minnows, brown bullheads, or black crappies are in a stream or river reach. So it is that we simply do not seek out such information, but we do proceed to compare water column contaminant concentrations against screening values for health protection, an unfortunate exercise dealt with in the next chapter. Why is this? There is a great curiosity about aquatic sites, for there are a great many occasions when comparative censusing *does* occur. Frequently macroinvertebrates are counted in sediment samples, but the merits of such exercises are not apparent; sites are not cleaned up to protect benthic macroinvertebrates. We also do not know how many fewer macroinvertebrates there need to be at a contaminated site (relative to a reference site) to signify an impact (presumably to the macroinvertebrates themselves). As with the misapplication of certain HQs (as in the earlier example of errantly thinking that a mouse HQ > 1 would affect owl health), we cannot apply a finding of a lesser site macroinvertebrate count to the health of other aquatic receptors sharing the same ecosystem. Finally, we should know better than to resort to conjecture to make a case that a site cleanup is needed. Thus we might find fewer macroinvertebrates at a site, and we might also discover that a preponderance of site macroinvertebrates are pollution-tolerant species. Before claiming that conclusive evidence exists to show that higher trophic species (e.g., fish) are health challenged, we need to know if indeed there are any fewer fish at the site. Could it be that pollution-tolerant macroinvertebrates are more nutritious than pollution-intolerant species? Could it be that pollution-tolerant species supply greater biomass to the ecosystem than pollution-intolerant forms?

It was the purpose of this chapter to have the reader understand that prudently, an ERA's most pivotal first step (after establishing that a site is indeed contaminated) is a screening exercise. The exercise would establish if any one animal contacts a site enough to develop a toxicological effect (of interest), and that a site harbors sufficient representatives of each species to drive the process altogether. This previously unknown screening could rightfully precede the most rudimentary of chemical screenings of the standing process (e.g., for chemical frequency of detection, for site-relatedness, etc.). Ostensibly, outcomes of "receptor presence-based screenings" could well serve to alert the ERA practitioner to the reality that, contaminated as a site might be, there could be no need whatsoever for an ERA (at least of the variety with which we are all familiar).

References

CH2M Hill (2001) Development of Terrestrial Exposure and Bioaccumulation Information for the Army Risk Assessment Modeling System (ARAMS). CH2M, Sacramento, CA.

DeYoung, R.W. & Miller, K.V. (2011) White-tailed deer behavior. In: Hewitt, D.G. (ed), *Biology and Management of White-tailed Deer*. CRC Press, New York.

Harestad, A.S. & Bunnell, F.L. (1979) Home range and body weight – a reevaluation. *Ecology* 6:389–402.

Tannenbaum, L.V. (2005a) A critical assessment of the ecological risk assessment process: A review of misapplied concepts. *Integrated Environmental Assessment and Management* 1:66–72.

Tannenbaum, L.V. (2005b) Two simple algorithms for refining mammalian receptor selection in ecological risk assessments. *Integrated Environmental Assessment and Management* 1:290–298.

Tannenbaum, L.V., Gulsby, W.D., Zobel, S.S., & Miller, K.V. (2013) Where deer roam: Chronic yet acute site exposures preclude ecological risk assessment. *Risk Analysis* 33:789–799.

US EPA (1989a) Risk assessment guidance for Superfund. Volume I: Human health evaluation manual (Part A), Interim Final. EPA/540/1–89/002. US Environmental Protection Agency, Washington, DC.

US EPA (1989b) Risk assessment guidance for Superfund. Volume II, Environmental Evaluation Manual, Interim Final. EPA/540–1–89/001. US Environmental Protection Agency, Washington, DC.

US EPA (1991) NPL Characterization Project; National Results. Office of Solid Waste and Emergency Response, Washington, DC: EPA/540/8–91/069. US Environmental Protection Agency, Washington, DC.

US EPA (1992a) Framework for Ecological Risk Assessment. Washington DC: Risk Assessment Forum. EPA/630/R-92/001. US Environmental Protection Agency, Washington, DC.

US EPA (1992b) ECO Update, Intermittent Bulletin Volume 1, Number 3, The Role of Natural Resource Trustees in the Superfund Process. Office of Solid Waste and Emergency Response, Publication 9345.0–05I.

US EPA (1993) Wildlife Exposure Factors Handbook, Volume I. EPA/600/R-93/187a. US Environmental Protection Agency, Washington, DC.

US EPA (1997) Ecological Risk Assessment Guidance for Superfund: Process for Designing and Conducting Ecological Risk Assessments, Interim Final. EPA/540-R-97–006. US Environmental Protection Agency, Washington, DC.

US EPA (1998) Guidelines for Ecological Risk Assessment. Washington DC: Risk Assessment Forum. EPA/630/R-85/002F. US Environmental Protection Agency, Washington, DC.

4 Toxicology and toxicity assessment in ERA revisited

ERA is admittedly in its relative infancy. It came into being less than 30 years ago, shortly after the US EPA introduced the famed Superfund program (Kendall & Smith 2003). Because of its young age, ERA and all health risk assessment for that matter, has been relegated to the status of a discipline, and at best perhaps, a quasi-science. Toxicology on the other hand, so often drawn upon to support what we term ERA, is without argument a *bona fide* science, "the science of poisons." Toxicology dates back to the earliest humans who used venoms and plant extracts for hunting, warfare, and assassination. As a formal science, it dates back to the late Middle Ages and the early insights of the physician–alchemist Paracelsus (1494–1541) who studied the human body's responses to chemicals (Gallo 1996). When the Superfund program identified the concern that plants and animals might be experiencing health effects from the chemical exposures they incur at certain industrial and other sites, a new avenue for applied toxicology was born. With centuries of knowledge behind it, in addition to honed capabilities, and great finesse, toxicology came to be viewed by many as ERA's guiding light and savior. The open-minded reader though, should be willing to hear that toxicology, for all the many centuries of experience it holds over ERA, hasn't risen to the occasion. In keeping with this book's practice of identifying alternative understandings of various ERA elements, and of sharing carefully considered insights and observations, there is much to offer after subjecting toxicology, in an ERA context, to its share of scrutiny. Unlike the previous chapter that concluded with a series of pronouncements that ERA practitioners would undoubtedly prefer to not have to hear and consider – because they could be so on-target – this chapter will lay out upfront its pronouncements concerning toxicology's place in ERA. A caveat or two first, though. The reader is reminded that the author is not a toxicologist. At the same time, the author is sufficiently fluent with the essential principles of toxicology to allow him to effectively apply his knowledge in this area to the ecological risk assessment phenomenon.[1] Free of the toxicologist's milieu,

[1] This is a reminder that we have learned already that there is no such thing as an ecological risk assessor.

Alternative Ecological Risk Assessment: An Innovative Approach to Understanding Ecological Assessments for Contaminated Sites, First Edition. Lawrence V. Tannenbaum.

and thus with no sense of inherent obligation to defend toxicology in its intended supporting role in ERA, it is only then that the author can try his hand at rendering what is hopefully some beneficial eye-opening. En route to such eye-opening, the reader would do well to consider that perhaps only a non-toxicologist can meaningfully review the role of toxicology in ERA, pointing out along the way where toxicology leaves us wanting for better science. This chapter will endeavor to show that:

toxicology provides little if any useful service to present-day ERA;
the present ERA paradigm has no need for toxicology anyway;
toxicologists working in support of ERA considerably overestimate their capabilities; and
the need to develop more ERA-attuned toxicological science is (apparently) not perceived by the regulatory powers-that-be.

The earlier highlighted distinction of the relative ages of ERA and toxicology deserves our attention. In the hundreds of years of development that the science of toxicology has enjoyed, its focus was solely on understanding how chemicals affect humans, be this in a context of occupational exposures or a context of assisting the medical field through the development of therapeutic drugs. It is also important to note that although toxicological study for these purposes may have frequently involved the use of non-human species as test subjects, such study was not directed at animal health proper. Additionally, where animal health *was* the focus, as in the development of veterinary science, this context is far removed from ERA's concerns over the health of non-domesticated species that have been exposed in the wild to contaminants that stem from anthropogenic sources. It is imperative to understand then, that it is only because of the creation of Superfund (in 1980!) and the ecological health concerns this program raised, that the field of toxicology turned its attention to an area it had previously not explored. Prior to ERA as we know it, toxicology had not dealt with the matter of estimating potential reproductive effects accruing to American woodcock from consuming PCB-laden earthworms at an electrical transformer dumping site. Similarly, toxicology had never been tasked with determining if a benthic macroinvertebrate species assemblage in a contaminated lake was sufficiently altered so as to impact lake productivity. Since ERA has a recurring and defining question (i.e., Are the chemically-exposed ecological receptors at a contaminated site likely to develop negative health effects?[2]), toxicology needed to adapt in order to render assistance to ERA. The case can be made that in 30 years' time, toxicology has not broken free of its traditional design and regimentation to do so. Let us see why.

[2] Chapters 8 and 9 will explain that this is really not the correct "defining" question of ERA.

Conventional toxicological research has the ability to render findings that have perhaps the least degree of associated uncertainty of any science. It is possible to design a study so that a singular variable is identified that accounts for any measured or observed differences between or among treatments. For other sciences and fields of endeavor, the opportunity to be this exacting and to be able to identify causation may not exist. A toxicologist for example, can conclude rather definitively that a drug added to an animal's diet is the sole cause of the excessive proliferation of a certain kind of cell within an internal organ, and the toxicologist can also know rather precisely what dose triggers such cell changes. The luxury of being able to access highly definitive cause-and-effect data can lead to the thinking that such data is appropriate for any and all applications. It is here though, that an unavoidable downside of toxicological practice, from the vantage point of ERA, presents itself. What makes for a high-quality toxicology study, as we have said, is the study's ability to control the testing environment such that the influence of a singular variable is known. Ironic as it may sound, it is this high level of control, most praiseworthy within the strict toxicology camp, which vastly strips away the utility of a study's findings for ERA purposes. In very real terms, ERA does not want to know how an inbred line of laboratory animals responds to an imposed chemical stressor, for this represents a significant departure from the actual biological (i.e., genetic) arrangement that exists at a given contaminated site (i.e., panmixis). In a similar way, ERA does not want to know how a laboratory test species responds to a chemical stressor that was administered over part of its lifetime, when site receptors today (other than occasional site immigrants) reflect tens of generations of site exposure. The non-parallel nature of the imposed chemical exposures of conventional laboratory-based, dose–response toxicity studies, and the exposures that ecological receptors receive in the wild, is beyond extreme. It is likely that toxicologists who sincerely want to assist ERA do not fully appreciate the radical differences of the two types of exposure settings. This book's purpose lies not in the castigation of toxicologists (or professionals of other environmental science disciplines) who support ERA. Nevertheless, before expanding on the subject of the ineffectual contributions of toxicology to ERA, it might be worthwhile to posit two simple reasons for toxicologists failing to see the sheer non-applicability of their work. It may first be that classically trained toxicologists are not able to break free of the specialized knowledge they possess because it is so ingrained in them. Thus, if they've learned that a mainstay toxicology design study element is that there is only a singular chemical stressor, they will lack an appreciation for the reality that ecological receptors at hazardous waste sites are rarely, if ever, exposed to a singular contaminant. Potentially, ego constitutes a second basis for the overlooked or ignored instances of incompatibility of the exposure settings of laboratory and field. When toxicologists stop to reflect on the extensive body of knowledge they possesses, where such phenomena as dose–response, chemical mode of action, and pharmacokinetics are often well understood, the possibility that the toxicological contribution tells us anything less than the whole ERA story cannot register.

To this point we have mentioned three rather universal features of toxicity study design that are responsible for furnishing information that ERA doesn't really need or value, namely dosing genetically constricted test subjects, conducting one-generation exposure testing, and testing chemicals one-by-one. The list of toxicity study design features that compromise the utility of the data that the studies generate, runs longer than this, though. Thus, ERA is not interested in the toxicological responses of animals exposed under the influence of controlled indoor laboratory conditions such as fixed-hour lighting, artificial (e.g., fluorescent) lighting, controlled room temperatures, and confinement to a cage or tank. Once again, these are conditions that do not coincide in the least with the arrangement in the wild. There are still more examples of how studies that ERA practitioners seek out are stripped of their utility. ERA does not want to know how test animals respond to a chemical form different from that found at the contaminated site of interest. The matter here is one of our failing to speciate the chemical form(s) occurring in soils, sediments, or surface water, but electing to use as critical evaluative tools the toxicity thresholds from laboratory studies that dosed test species with one specific commercially available chemical form. ERA also isn't interested in the toxicological responses that follow from chemicals having been administered through some artificial route (e.g., introperitoneal injection). We are (or should be) prompted to ask why ERA utilizes toxicity information drawn from studies that in such pronounced fashion deviate from the conditions of real-world chemical exposure. It has already been suggested that ERA practitioners might be oblivious to the disconnects of laboratory and field exposure settings. Perhaps we should give ERA practitioners more credit than this, but the uncertainty sections of remedial investigations (RIs) and ERAs bear out the seeming veracity of the suggestion. These sections consistently fail to mention any of the thus-far listed shortcomings that bear on study design. Instead, the uncertainty sections acknowledge the more basic and generic points that bear more on hazard computation. Standard fare in uncertainty reporting extends to such things as acknowledging that larger sample sizes for various media (e.g., sediment) lead to the computation of more accurate exposure point concentrations, and that high dose-to-low dose extrapolation may not be linear in nature.

We have yet to discuss what is potentially the most pervasive and damaging design feature of toxicity study in ERA. Not surprisingly, this feature too is not known to be addressed in the uncertainty sections of ERAs, and the author would go so far as to intimate that this feature has never been acknowledged in these critical textual treatments. It is a fact that the very organisms that toxicologists expose to chemicals have had no prior chemical exposures. While this design arrangement is the toxicologist's choice, the ramifications of the arrangement lead to the generation of dose–response information that is possibly the least utilitarian of all. Let us couch this design feature of critical toxicity study in the phraseology employed in the earlier paragraphs. ERA is not interested in learning how previously non-exposed organisms respond when first confronted with chemical exposure, be this in the form of an administered and internalized (chemical) dose, or through having contacted a contaminated environmental medium. While some might elect to disagree, we have established in the previous chapter that ERA intends to understand the health risks that accrue to receptors that are presently, and have already been (for quite some time) chemically exposed precisely because they live amidst site contamination. Further, and as we have already noted, we are aware of many instances of site receptors lacking the complete ability to free themselves of their site exposures by physically moving away from them. We need only summon forth images of a waterbody's sessile benthic forms (e.g., certain macroinvertebrates), or of earthworms in a managed site's operable unit, to cement the concept. Benthic macroinvertebrates (e.g., amphipod or copepod crustaceans) in a lake portion do not migrate hundreds and thousands of feet to find themselves, as first-generation representatives situated atop, or somewhat burrowed into, contaminated sediments at a given site. Similarly, earthworms are extremely constrained in their lateral movements within the soil column. As with the crustaceans we encounter in waterbody sediments, earthworms that we encounter today at a land parcel, are probably situated only tens of meters from locations that were occupied by countless generations before them. Similar to the suggestion in the previous chapter, we do not suspect that there are individuals venturing out to contaminated lakes and streams to secretly release batches of otherwise healthy laboratory-reared crustaceans in an attempt to boost populations that they fear are depauperate. Were such secretive work to be occurring though, we might then (and only then) have just cause to access the conventional toxicity studies that subject previously unexposed organisms to chemical stressors.

If our concern is with the health of ecological receptors that have been exposed for a long time to chemical stressors, studies that employ test subjects that have had no prior chemical exposure have little to offer us. The reader would do well to imagine the following experimental arrangement. Fish from a mercury-contaminated lake are collected and transferred to laboratory aquaria where the column water and sediment are spiked with mercury of the same form and concentration as that of the lake from which the fish were taken. Additional matching aquaria are set up in the laboratory, but into this second group of aquaria are placed fish of the same species and size as were harvested from the lake, but that were commercially raised in a chemical-free environment. Would we not expect the commercially reared fish to present with toxicological illness more often and sooner than the lake fish? Indeed we should, for encountering a toxicological stressor for the first time is a situation that is surely wont to elicit pronounced effects. If we had to wager, we should anticipate the lake fish managing extremely well in their aquaria, this because they have undoubtedly adapted over months, years, and generations to living in a mercury-contaminated environment. Let us now imagine the following applied toxicological approach to address a concern over potential health effects occurring in fish populating a mercury-contaminated lake. Groups of commercially, laboratory-reared fish of a common test species (one that does not happen to occur in the lake) are placed into a series of aquaria with variable mercury concentrations. It becomes quite evident that a mercury concentration of x in the aquarium water is the EC_{50}. If the mean column-water mercury concentration in the lake of interest is $2x$, would it be fair to conclude that the lake fish are imperiled? Many well-meaning ecological risk assessors would respond in the affirmative, and herein lies the damage done to our assessment process when we draw on toxicity studies that are not truly applicable to the situation at hand. The immediate problem is that of our failure to recognize when toxicity studies are simply not applicable. In the example here, the non-applicability issue has nothing to do (necessarily) with the test species being different from those found in the lake. To the contrary, it could well be that because a more generalist fish species was used as an (excellent) surrogate for a range of other fish species occurring in the wild, the ability to extrapolate from that species' response may be enormous.

In our aquatic example above, the non-applicability of the toxicity study derives of course, from having employed fish with no prior mercury exposure to represent the fish presently occupying the lake. The reader should rest assured that for this hypothetical case, and for any actual contaminated site's assessment of ecological receptor health, we are not suggesting that all toxicity studies need to be redone; that from this point forward, we need to replace unexposed test subjects with pre-conditioned ones. One can only imagine the technical complications involved with implementing such an *avant garde* initiative, anyway. What concentration(s) of a chemical should be used? How long should the pre-conditioning period be? And so on. A first pivotal point to secure here is that the concept of using pre-exposed or pre-conditioned test subjects would unquestionably allow for apples and apples comparisons, something we fall far short of having when we apply any of our existing studies. Although we are not suggesting that a new class of toxicity study be born, the merit and potential gains from having such studies at our disposal should be recognized and appreciated. Another pivotal point asks why we have never heard of the prospect of using prior-exposed test subjects in screening or assessing sites, given that the chemical exposures that ecological receptors experience are almost always not new to them. Surely our science acumen is amply robust to out-muscle the challenge of bringing about the new wave of toxicity studies. With some fortitude, we could surely identify several species of dinoflagellate, insect, shrimp, fish, earthworm, and probably rodent and bird as well, that are amenable to being maintained, through diet or otherwise, in a contaminated state. We might suggest that anticipated complications associated with establishing a standardized prior-exposure model stands as the reason for our never having broached the concept of testing with pre-conditioned animals. Thus, standard test species (e.g., *Eisenia* spp., *Hyalella* sp.) live but so long, and culturing intended-to-be test animals for a goodly portion of their earlier life, would likely rob them of their vibrancy. The poor viability of such chemically reared test animals in any subsequent toxicity studies might simply be a reflection of the advanced age of the test subjects at the time when they are called upon for use. Similarly, poor and erratic responses might not stem from chemical stress encountered during the testing, but might instead derive from some weakened status traceable to the pre-conditioning stage. We might like to believe that what accounts for our never having heard of the concept of testing with pre-exposed or pre-conditioned subjects, are the technical considerations of toxicologists, some of which we have explored here. The author however, would like to suggest that there is a more simplistic reason for the fully non-existent concept, namely that no one has ever thought of it. Alternatively said, no one ever came upon the idea because no one recognized that ERA's status quo approach to toxicological screening is so severely plagued. No one ever sat down to consider the lacking utility of a comparison of the toxicological response of a previously non-exposed organism to a comparable organism living in the thick of a chemical contamination site. In the earlier example of lake fish reputedly being continuously exposed to a mercury concentration double the EC_{50}, are we to conclude that the lake fish are without question presenting with the effect being monitored? What would we say if pursuant to the EC_{50} comparison, we examined lake fish and found them to be without effect? Would we say that the EC_{50} was correct and had served us well?

The toxicologists who contribute to the ERA process incriminate themselves in two ways vis-à-vis this discussion topic. As they have for two decades or more already, they continue to comb the literature for any and all studies that subjected reasonable test species to chemicals that occur at contaminated sites. Ignoring the drawbacks that plague the studies (enumerated in the foregoing pages), the toxicologists see as utilitarian their efforts to utilize the reported information in the published papers to establish no effect-level or effect-level chemical doses to be used in various assessment tasks. Such efforts are hardly utilitarian though, and of all players in the ERA game, the toxicologists should recognize why. Ninety-five percent of the TRVs regularly used in ERAs derive from studies that predate the earliest of ERA guidance, and 45% of the studies were conducted before the creation of the US EPA, whose Superfund program put ERA on the map altogether (Tannenbaum 2001). Although perfectly legitimate TRVs can be derived from a study no matter how long ago it might have been conducted, a study that predates ERA tells us unquestionably that it was not designed with ERA thinking in mind. Not surprisingly, the historical literature so often drawn on by toxicologists only reports the degree of absolute difference observed between the exposed or treated group and the control group for a given measured parameter. Thus, we might know that exposed rats ended up weighing 10% less than did controls, but the manifestation of such weight loss is unknown. Presumably ERA would want to know if animals that lost weight through their chemical exposures have significantly shorter lifespans, are less active, mate less often, or fail to reproduce as well as controls. If a truly utilitarian study could be conducted in a laboratory setting, at a minimum it would maintain the animals well beyond the dosing phase such that longevity and reproductive performance could be assessed. We can well understand why the toxicological studies of yesteryear didn't do such things; the studies weren't oriented in the least to the ERA field because ERA hadn't yet emerged. The toxicology studies of yesteryear accomplished perfectly well what they were intended to do, though. They indicated such things as the specific organs and enzyme-mediated biochemical processes that are sensitive to given chemicals, and the shape of the dose–response curve for the specific set of imposed test conditions.

The utilization of toxicological response information that derives from classical experimentation (i.e., where no focus whatsoever was placed on ERA) only serves to highlight the extensive nature of data gaps in our young discipline. ERA practitioners generally understand the concept of a data gap to refer to a chemical for which no readily accessible toxicity information exists (altogether, or for a specific route of chemical uptake). In truth however, every study-derived TRV brings along with it another costly data gap, and one that the rank and file undoubtedly overlooks. This can become evident only if we put certain notions out of our thinking. We must temporarily ignore the rather extensive list of drawbacks of the available laboratory studies upon which we seize (e.g., use of inbred animals) and also the fact that the studies we draw on do not evaluate the endpoints that are most meaningful to us (e.g., longevity, reproductive output). It goes without saying that we must too ignore the overarching difficulty of HQs that stem from the use of derived TRVs that are not expressions of risk. With these great many considerations peeled away, we are free to ask if the dose–response information that the studies do report (e.g., 10% weight loss in treated animals) speaks to toxicological effects as we would want to know them. Is 10% weight loss resulting from a certain dosing regimen detrimental to an animal or to an entire population? If we believe it is detrimental, what allows (some of) us to espouse such an opinion? To be fair, it could be that test animals that lost weight through their dosing regimen actually emerged healthier than when they entered the study, and healthier too than the study's control animals. This leads us to a discussion of deleterious effects and specifically how we go about identifying them.

In classic and conventional toxicology studies, we almost always begin with the null hypothesis that a dosed population will react no differently from a control one, and our intention is to disprove this. Thus we would like to find that at a sufficiently high dose, a chemical or drug laced into the diet, applied topically, or infused into the air, triggers cells to respond in a manner noticeably different from that which occurs in non-exposed animals. The standard adopted by the scientific community for claiming to have identified a noticeable difference is a statistical one. If we can be 95% sure that an effect is real (i.e., that the effect we've latched onto is not simply due to chance), we have the necessary backing to verify the discovery. In very real terms, a peer-reviewed scientific journal will be much more receptive to a manuscript that found a statistical difference than it will be to a manuscript that reports that no difference was found between or among test groups. There is a great divide though, between meeting a peer-reviewed journal's standard for publication acceptability, and having in-hand a toxicological benchmark that can be reliably used to assess the potential for health impacts to accrue in ecological receptors exposed to chemicals.

The follicle density test (FDT), developed in the late 1990s by the US Army Center for Field-based Ecological Risk Assessment Method Development can shed a great deal of light on this discussion that revolves around statistical significance in toxicological outcomes and the designation of outcomes as deleterious. Among the method's accolades is the highly distinctive FDT guidance document cover (Fig. 4.1), recognized as likely being the most descriptive of any other comparable techniques-type guidance. The FDT understands that exposure to chemicals, either singly or in combination, can cause mammals to lose their hair or fur. With such losses, precious body heat can be lost in the winter months, thereby posing thermoregulatory challenges to animals. The FDT calls for collecting small rodents at contaminated terrestrial sites and nearby habitat-matched reference locations, and shaving off a 1 cm^2 patch of fur on each small rodent's rump. The number of hair shafts (follicles) is counted for each animal, and the site and reference location population means are compared with a Student's *t*-test in support of a determination. Table 4.1 provides summary information for an Army installation that ran the FDT on a number of occasions for one of its sites. Of note, there were certain years when the FDT was not conducted although the site was still very much in the analysis phase. The gaps reflect nothing more than budgetary constraints at the installation's environmental office that stood in the way of more data collection. As the table indicates, there are two dominant small rodent species that regularly occur at the site, consistent with what had been reported in

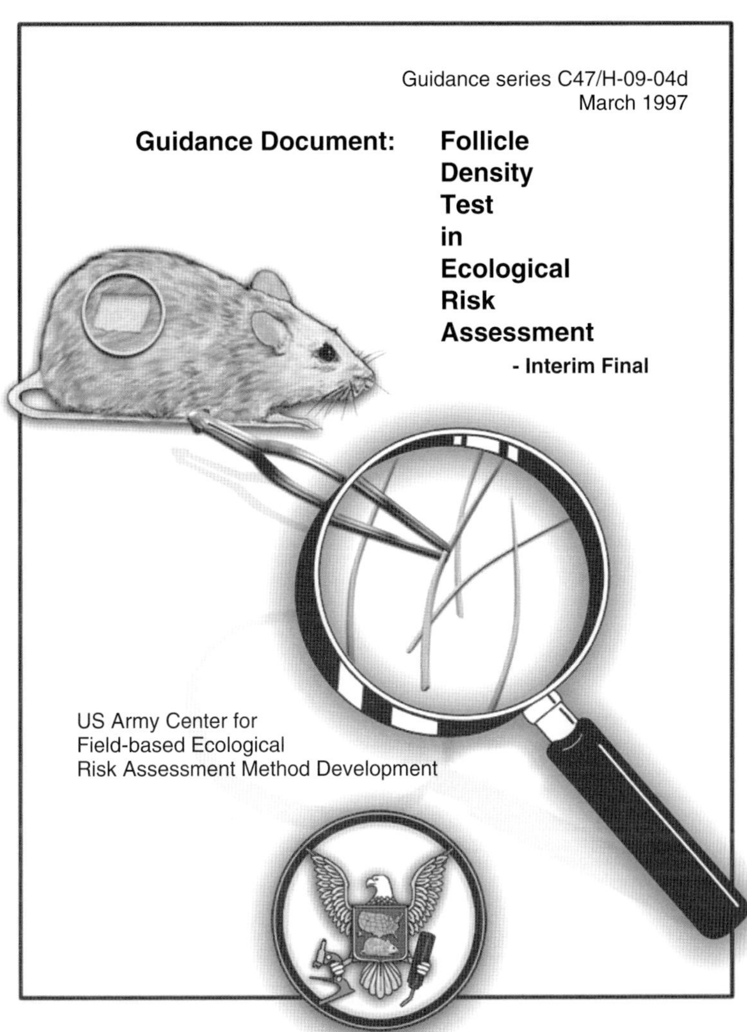

Fig. 4.1 A most self-explanatory guidance document cover, the one for the Follicle Density Test for rodents.

biological surveys for the installation from several decades earlier. Several other occasional small rodent species were captured at the site and reference location, but their limited numbers preempted them from submitting to FDT analysis. Several species-specific differences are evident in the data. The first, relegated to simple descriptive biology, is that meadow voles (*Microtus pennsylvanicus*) have less densely packed follicles than do white-footed mice (*Peromyscus leucopus*).

Table 4.1 Cumulative FDT results in female small rodent species at Fort ________, WI.

Test species	Year	Mean follicle density (follicles/cm^2)		% decrease in follicle density at site	Was decrease statistically significant? ($p < 0.05$) (Y/N)	Was thermo-regulation compromised? (Y/N)
		Reference location	Contaminated site			
Meadow vole	1999	91 ($n = 21$)	85 (20)	6.6	Y	
	2003	102 ($n = 20$)	90 ($n = 24$)	11.8	Y	
	2005	79 ($n = 26$)	74 ($n = 18$)	6.3	Y	
White-footed mouse	1999	133 ($n = 23$)	68 ($n = 23$)	48.9	Y	
	2002	151 ($n = 22$)	111 ($n = 19$)	26.5	Y	
	2003	111 ($n = 21$)	80 ($n = 26$)	27.9	Y	
	2004	124 ($n = 25$)	81 ($n = 20$)	34.7	N	

Two other differences bear on the applied FDT; in each of the seven comparisons, the follicle count is less at the contaminated site relative to the reference location; for all the comparisons, the magnitude of the observed follicle count decrease was noticeably greater in the white-footed mice. Notably, all but one of the seven comparisons in the table was statistically significant. At this point the reader is asked to interpret the FDT findings in the table's last column.

Endeavoring to follow through with the interpretation exercise is fraught with great difficulties. The least of these is being misled into thinking that because statistical significance was observed in a great many cases, this finding confers that effects related to follicle losses have taken hold. Although all but one comparison was statistically significant, we do not know that any animals suffer from compromised thermoregulation. With no more discussion than this, every empty table cell in the last column should bear an "N". As already mentioned, findings of statistical significance only tell us that it is not due to chance that test subjects respond in a particular way (here, that small rodents from the contaminated site present with relatively fewer follicles/cm^2). What *is* known, even with the minimal data supplied in the table, is that some chemical influence at the site is causing follicle reductions.[3] A next difficulty regarding data interpretation that could be labeled severe is that of concluding that site rodents are health-compromised (presumably in their thermoregulation) because of the magnitude of the follicle losses that were not in the direction of favorability. This difficulty is certainly applicable to the white-footed mouse, where the smallest net reduction was >25%. Deeply ingrained in the thinking of many an ERA practitioner is that changes of 20% that are not in the direction of favorability are indicative of impact. The 20% threshold figure is loosely rooted in a data review of sediment toxicity test outcomes. At best, the bandied-about 20% figure is thought to reflect that difference in a parameter of interest (e.g., enzyme level, organ weight, bone length) that is beyond a given study system's background noise level (i.e., the naturally-occurring variability or scatter that exists in virtually all datasets). The reader is reminded that Chapter 9 pays a great attention to the so-called "20% rule" that is so often invoked by regulators and other stakeholders in site-specific ERA doings.

The difficulties with Table 4.1 thus far discussed pale in comparison to a most egregious two-part problem with the table and the FDT itself. The design of the FDT may have been well-intended, but the fact remains that we have no basis for thinking a lesser follicle density to be a harbinger of compromised thermoregulation. It could very well be that lesser follicle densities that arise in response to chemical exposures seriously challenge mammalian ecological receptors to maintain a constant core body temperature. Until such is established though, the basis of the test is only conjectural in nature. As a consequence, there is no validity to conducting FDT studies altogether, a blow perhaps to the developers of the test and those who historically have elected to use it. Even if the FDT essential premise *was* wholly valid (i.e., a reduced follicle density is in fact, an indication that thermoregulation is compromised), we would still be faced with critically absent science. For the test to have any utility, we would need to know how much of a reduced follicle density signifies some degree of thermoregulatory impairment.

This FDT review is not as libelous as it would appear. In truth, there is no such test, nor is there such an agency as the US Army Center for Field-based Ecological Risk Assessment Method Development. For the reader with a highly perceptive and scrutinizing eye, the *giveaway* was the unlikely logo of the fictitious agency credited with developing the test, namely an eagle clutching a microscope in one set of talons and an inverted rodent cage water bottle in the other. The FDT case nevertheless sets the stage for a consideration of the laboratory testing data that is routinely accessed in present day HQ-based ERA.

[3] The underlying assumption with the FDT is that where a comprehensive list of habitat features are shared by the two rodent trapping sites (e.g., soil type, vegetative cover, etc.), identifiable follicle pattern differences in site rodents necessarily stem from the site soil's chemical profile.

The far left column of Table 4.2 lists a range of predominantly somatic effects elicited in a great number of toxicological studies that have been used to form the basis of many TRVs. Potentially each toxicological effect can compromise an animal's health in some way. Presumably though, for most if not all of the effects listed, ERA practitioners select the TRVs they do, because of an assumed or demonstrated linkage to the mainstay toxicological endpoints that dominate ERA concerns, i.e., those identified in the remaining column headings. [We should not deny though, the possibility that oversized livers induced by chemical exposure

Table 4.2 Biological significance fill-in exercise.

Toxicological Effect in test species (bird or rodent)	Degree of change in toxicological effect in treated animals (percent change relative to control animal response) that signifies compromise of health compromise			
	Survival/ longevity	Reproduction	Growth	Behavior
Oversized liver				
Undersized liver				
Microscopic hepatic lesions				
Oversized liver				
Undersized liver				
Tubular degeneration (kidney)				
Reduced thymus weight				
Body weight loss				
Reduced food consumption				
Increased number of hyaline droplets				
Reduced growth				
Delayed bone ossification (femur)				
Reduced birth weight				
Hypertension				
Behavioral abnormalities				
Blood effects				

testing are harmful in and of themselves, as in critically compromising hepatic circulation. Where such assertions are harder to prove, it is not uncommon to hear a wishful but unproven argument tacked on, all in the interest of furthering the use of such studies. As the argument goes, animals that are already health-challenged (say, because they bear an oversized or undersized internal organ) are less fit to withstand the next environmental stressor that might come along (say, a particularly harsh weather pattern spanning two or more successive summers).[4]] More than likely, using the oversized liver outcome of the toxicological study as an example, the ERA practitioner is concerned that endpoints such as longevity and growth will be reached.

It is unlikely that ERA practitioners in the main are fluent with that degree of organ or organ system compromise that challenges and threatens the essential function of these tissues. Thus we might know what chemical and how much exposure to it causes reduced thymus weight, but we probably don't know by how much a thymus weight needs be reduced to produce a dysfunctional thymus. It is therefore a weak argument to say that ecological receptor health is unacceptable at a site because there are HQs greater than 1.0 for reduced thymus weight. It is a still weaker argument to say that a site needs a cleanup because site receptors (we suspect) are saddled with reduced thymus weights when we are fully unaware of even a single receptor at a site of concern bearing such an effect. We can probably live without our knowing the tipping point for reduced thymus weight signifying impaired thymus function. We recall that chemical exposure testing was never conducted to ascertain when a thymus is too small to serve out its proper biochemical roles. That we are able to calculate a HQ for reduced thymus weight merely reflects that a literature review for one chemical (or perhaps a few) turned up a sufficiently robust study to allow for TRV generation with reduced thymus weight as the endpoint.

ERA practitioners should be much more fluent with the observed effects of animal testing as these relate to the mainstay toxicological endpoints (survival, reproduction, etc.). They should be able to populate the blank cells in Table 4.2 with approximate percentages. The exercise that Table 4.2 fosters keeps us mindful that ERA intends to track deleterious effects. An oversized or undersized internal organ is not itself deleterious, although there may be exceptions to this. (Thus, growth being compromised because of observed reduced growth, or behavior being compromised because of observed behavioral abnormalities are clumsy considerations. This explains the two hatched cells in the table that needn't be dealt with for the fill-in exercise.) Organ size *is* deleterious when it is

[4] The weakness of such arguments is evident at historically-contaminated sites. We consistently find these to be fully populated (i.e., as populated as reference locations), and the decades that have elapsed have surely provided ample opportunity for 'next environmental stressors' to have materialized. We cannot wait indefinitely for theorized next environmental stressors to show up.

skewed in one direction or the other such that an essential biological function is compromised or completely impeded. The reader could try his/her hand at filling in the table, but it would be unfair to again present a deception. For none of the listed effects of the table's first column do we know how much deviation from the norm compromises the effects we have labeled as 'mainstays'. To those who would challenge this point, and argue that there is utility nevertheless in assuming the task of filling in the table, there are imperative considerations that cannot help but quell such thinking. To begin with, normative ERA practice is to compute HQs for such endpoints as hypertension, increased hyaline droplet number, and blood effects. But why?[5] It is certainly not thinkable that a site would ever proceed to a Record of Decision, wherein it is explained that the preferred remedial action (e.g., soil removal and replacement with clean fill) is intended to restore normal numbers of hyaline droplets to future populations of site receptors (perhaps shrews and squirrels). If we are bothering to compute HQs for potential hyaline droplet number increases or for hypertension, and there is insufficient environmental agency muscle behind unacceptable HQs for these endpoints to go forward with remedial action recommendations, we need to get more in touch with what we routinely do. In truth, the reason we would calculate a number-based HQ for hyaline droplets is only because in our limited bag of tricks (i.e., our ability to access available compilations of toxicological benchmarks for wildlife; Sample et al. 1996; US ACHPPM 2000), we had the good fortune to find a no-observed-adverse-effects-level (NOAEL)- or lowest-observed-adverse-effects-level (LOAEL)-based TRV for a given chemical. Another imperative consideration: it is most prudent to note that we will never know that our site animals actually present with the toxicological conditions that HQs >1.0 suggest are taking hold. Thus, we will never collect rodents from the field and measure their blood pressures, although realistically such readings could be taken. After this chapter's earlier review of the drawbacks of standardized chemical exposure studies, we could only in earnest, attempt to sell the argument that skunk or mink are at risk of hypertension from their site chemical exposures if we furnished the somatic data to support such claims. Short of doing so, we are left with the irresolvable challenge of finding utility with HQs that are so unconvincing.

There are still more imperative considerations. One of these would present itself if we were, in fact, to invest our time and energy to measure those somatic features, like blood pressure or hylaline droplet count, that are not readily observable from the animal's exterior appearance. If we were to find statistically significant differences in site receptors (relative to receptors of the

[5] The weakness of such arguments is evident at historically-contaminated sites. We consistently find these to be fully populated (i.e., as populated as reference locations), and the decades that have elapsed have surely provided ample opportunity for 'next environmental stressors' to have materialized. We cannot wait indefinitely for theorized next environmental stressors to show up.

same species at an appropriate reference location) that were not in the direction of favorability, when should we imagine the differences first arose? We recall that most contaminated sites that submit to ERAs are 30 or more years old, and that the generation times of ecological receptors are vastly shorter than man's. In the earlier FDT example, it would not be fair to suggest that reduced follicle counts developed for the first time in 1999, the year the installation first learned of the test, became enthralled with it, and had the available finances to apply it. Reasonableness is the very heart of health risk assessment. To suggest that each and every instance of a biological measure at a contaminated site being found to be different from the norm, first arose in the very year of the discovery of the skewed measure, is shameful. It is shameful when a more reasonable toxicological interpretation could be that the difference occurred a decade or more earlier, with the population having demonstrated since then that it can well tolerate the change that sprang up. It is most unfortunate to have to anticipate the knee-jerk interpretations of toxicologists and regulators regarding observed differences in field-collected specimens. It is particularly unfortunate when these parties conclude that a problem exists at a contaminated site when the data they reviewed en route to the interpretations stem from newly developed testing schemes. Novel biological measures and the great efforts invested to bring them online are likely to be met only with resistance and rejection. The more often it happens that new ideas are shot down for the very fact that they are represent a departure from the standard fare, the more reluctant others will be to try their hand at bringing new assessment tools online that could improve the ERA process.[6]

Every biological difference is not calamitous for a population or for a site's ecology. An unstated expectation of regulators though, is that each and every biological measure must highly or perfectly match at a site and its associated reference location. Further, until such matching is found to occur or is explained to the regulator's liking, the contaminated site must remain under the watchful eye of the regulator and be subject to the regulator's demands for ongoing study. Such an arrangement longs for toxicology to weigh in to correct this errant thinking, but the case can be made that toxicology hasn't asserted itself to supply the needed support. We can cite other examples of toxicology's shortcomings that are quite apparent and where we have no reason to believe that there might be efforts afoot to rectify the weaknesses.

- Regrettably, toxicology in ERA remains anthrocentric; toxicologists impose their (human) thinking and the conclusions they draw onto the animals we study. In short, if we determine there is a chemical exposure-induced

[6] Bias is discussed Chapter 7. Chapter 9 reviews a novel, fully developed, and vetted field-based biological measure testing scheme for terrestrial sites, and discusses the prospects for its adoption by regulatory and other agencies.

effect, we conclude that the organisms that display it are more than aware of it. In conducting newer studies and in interpreting others of yesteryear, toxicologists facilitate the thinking that dosed organisms that develop effects as a consequence, to even include changes that are not externally visible, appear differently to their conspecifics, congenerics, and all other elements of their local ecology. Notably, the thinking extends to the detriment of the changed animals themselves. The thinking is best captured in Fig. 4.2, where reproduction

Fig. 4.2 Anthrocentric thinking in toxicology assessment as it assists ERA. We might know of the chemical-mediated changes occurring to site receptors, but do the receptors know of them?

stands to be compromised. In practical terms, anthrocentric thinking amounts to projecting onto another organism the capability to recognize visually or detect in some other fashion, a changed somatic feature in a site mate, and further, to respond or react differently towards that chemically affected organism. Non-human receptors have been described as having exceptional intellect, and have been documented with uncanny powers of detection, as in dogs being able to identify humans with certain cancerous cells (Sonoda et al. 2010). Yet for the juvenile field of ERA that is heavily rooted in the practice of repeatedly drawing on a small universe of toxicity studies, it is beyond the toxicologist's right to assume that a measureable effect is apparent to the organism that bears it, or to the other organisms with which the 'affected' one interacts. So then, it might be that *the toxicologist* is bothered by the discovery that an organism is seemingly behaviorally or somatically short-changed because of the chemical exposures it has incurred (in the lab or in the field), but this sentiment should not be allowed to spill over to the ecological dynamics of the contaminated site of interest. We must question the right of the toxicologist to seize on, and become bent all out of shape over a discovered liver weight increase in the field mice of a contaminated site. Why must we be bothered about such an effect when we can't be sure that the effect is even registering on the proverbial radar screen of the field mouse? ERA is not here to make the ecological risk assessor feel better; it's here to ameliorate the condition of the ecological receptor and the site ecology should it have been impinged upon. To do things correctly though, we need to establish that there's been an impingement. Thus to get more mileage out of an observed chemically induced effect, toxicologists have their work cut out for them. Research is first needed to show – borrowing from Figure 4.1 – that organisms with plussed-up hyaline droplet numbers behave unusually, or experience negative consequences in one or more biological functions. Until this occurs, all other value judgments are valid, i.e., those other than the one that says that an excess hyaline droplet count is unhealthful. Perhaps female Hispid cotton rats *can* well detect excesses of hyaline droplets in conspecific males, and perhaps too, the females empathize with these males because they know from textbook reading, how hard it is for them to manage a stigmatizing feature. Maybe because of their empathy, the females go out of their way to differentially mate with hyaline droplet-challenged males, and reproduction is boosted as a consequence. Maybe then too, a chemically induced increase in hyaline droplets is a good thing! – or are *we* now guilty, like so many ecotoxicologists, of imposed anthrocentric thinking? There are ways to extricate ourselves from the imposed anthrocentric-thinking predicament. A first step in that extrication process is to acknowledge how ill-prepared we are to assess ecological receptor health even though we find ourselves armed with a toxicity study or two. To borrow on our hyaline droplets again, do we know that an excess of them actually compromises function of the organ in which they are found? Do we know that because a controlled laboratory study produced an increase in hyaline droplets that the same effect arises in the wild (at the contaminated site of interest)? And even before all this – for the non-toxicologist ERA practitioner who might be advocating for a site cleanup precisely because HQs were greater than 1.0 for excess hyaline droplet formation – does he or she know what organ is home to hyaline droplets? (For the reader who doesn't know, it is the kidney.)

- The earlier toxicology study-oriented exercises of this chapter, fabricated or actual, and so much of the foregoing text, is calling out for a distinction of paramount importance to resonate throughout the ERA camp. Thus we may speak of *first-order toxicology* and *second-order toxicology*, indispensable new terms for the field. *First-order toxicology* refers to the information gleaned from cautious high-quality research that identifies a chemical in its capacity to act as a poison. First-order toxicology principally applies to empirical research, be it on the order of a Master's thesis or PhD dissertation, or to the applied animal dosing studies with which we are so familiar. Notably, first-order toxicology establishes considerably more than the simple identification of a chemical as a toxin. Although not an exhaustive list, it also establishes:
  - applicable routes of chemical uptake;
  - the specific tissue(s) that the chemical disarranges;
  - the conditions or diseases that may ensue from the tissue effects;
  - thresholds-for-effect/dose–response;
  - rates of metabolic breakdown of the chemical (including the possibility of first-pass metabolism);
  - chemical sequestration in certain body compartments; and
  - the reversibility of effects pursuant to the chemical stressor being removed from the test organism.

All of the list items have some utility in ERA. Conspicuously absent from the extensive list though, is a critical informational piece, and one without which ERA cannot produce meaningful results. What we might term second-order toxicology is that capability to articulate the degree of chemically posed change that signifies health impact. Remarkably, for virtually all the studies we might access to assist with HQ calculation, we do not know this all-important delta. Let us begin with hyaline droplets, and proceed from there to fully elucidate the concept of lacking second-order toxicology in ERA, and the toll that this places on our field of endeavor.

The proliferation of a given cell type brought on by a chemical exposure is a sign that the body's capacity to regulate itself in a certain way has been overcome. Too many hyaline droplets in a kidney could potentially compromise kidney function because of the space they occupy, the droplets interfering with the activities of other cell types within the kidney. For one reason or another, the kidney might become less efficient in voiding waste through urine production because of the excess number of droplets. Alternatively, the excess droplets might cause the kidney to grow larger overall within the abdominal cavity. The enlarged kidney might then garner a greater blood circulation than it had previously, and be responsible for less than adequate blood perfusion occurring at other nearby organs. It may be that because of the energy channeled into having a certain type of cell proliferate, the energy demands for one or several other tissues failed to be met. As a consequence, short-shrifted cells and organs may have adopted a status of compromised function. The central point is that an excess of hyaline droplets brought on by chemical exposure is potentially detrimental to an animal, and it would be naive to think otherwise. If we are of a mind to believe than an excess number of droplets could cause an animal to move about sluggishly (and thereby be at greater risk of being predated upon) or retain ammonia and other wastes that should ordinarily be voided (thereby increasing risk of infection and death stemming from it), we have an acute need to know how much of an excess of hyaline droplets spells danger. We have no such information, however, and this begs the question of what utility, if any, is there in tapping into toxicity studies that only document the phenomenon of droplet increase, and relate how much of an increase was observed. If dietary exposure to Chemical X elicits a statistically significant 15% increase in the hyaline droplet count, HQs greater than 1.0 for Chemical X are at best, informing us that the receptor is consuming enough of Chemical X to increase its hyaline droplet count by 15%. Is an animal with a 15% increase in hyaline droplets health challenged, however?

Earlier, the prospect of conducting toxicity tests on chemically *pre-exposed* organisms was proffered. Realistically, we will never witness such a testing scheme materializing; the concept will be perceived as being either too abstract or too difficult to execute. The concept deviating so extremely from the status quo approach stands to be the greatest impediment to exploring a new toxicity study model. Although we can understand human nature, the fact remains that we are left without the toxicity studies we need. The studies that we *can* access do not incorporate the reality that ecological receptors live with contamination and are not always being exposed to it anew. In a similar way, it might sound unfair to ask that for each potential chemical we might come across at contaminated sites (and for which we presently have first-order toxicology-based TRVs only), we conduct altogether new studies that are expressly designed to uncover the deltas we so desperately need. Human nature being what it is, we should anticipate a great disinclination to acknowledge that the tools we have commonly employed to date in ERA are ineffective. The notion of conducting hundreds of tests with unusual and highly variable endpoints, and often including longevity as one endpoint to be monitored, would not be well received to say the least. Table 4.3 provides several examples of needed second-order toxicology information (i.e., the deltas), and how we could go about securing it. The reader will note that initiating the table is a now-familiar example.

The utility of the Table 4.3 exercise becomes evident when endeavoring to populate the table's third column. The ecological risk assessor who would have no qualms in calculating a HQ upon finding an available TRV, must now come to grips with why the effect around which a NOAEL- or LOAEL-based TRV is constructed, is deleterious. (The reader will note the usage of 'presumably' or 'seemingly' in the third column entries.) What's the trouble if a liver is oversized? What's the worry if laboratory-dosed rats did not gain as much weight as did controls? If the ecological risk assessor's 'trouble' or 'worry' is only that he/she would have preferred that there had been no differences between treated animals and controls, or between site and reference animals, there is no need to complete the table's last columns. If we've discovered that we don't really know that an effect is deleterious, we should not pursue the second-order toxicology studies described earlier. Moreover, if HQs greater than 1.0 were computed in the absence of second-order toxicology information, the ecological risk assessor must at this time recognize the non-essentiality of the computations.

Several pages back in the discussion it was stated that the toxicologists that contribute to ERAs incriminate themselves in two ways. The first of these was that they favor the use of studies that can quite easily be demonstrated to be lacking. The statistically significant differences of study outcomes (e.g., a 15% reduction in kidney weight) are interpreted to be biologically significant when it is not known if such is true. Because of that highly suspect interpretation, those in the ERA field proceed to calculate HQs, assuming all along that the study outcome was an adverse one, or perhaps not even stopping to consider if an outcome was adverse. Realistically, a 15% weight-reduced kidney might be highly functional, and to be fair about it, such a kidney might actually outperform a heavier one. (Perhaps evolutionarily, kidneys were smaller and as a side-effect of the successive inbreeding by common laboratory animal suppliers, man has acted to inadvertently 'manufacture', as it were, still healthy specimens that happen to present with larger kidneys. Should a toxicology study happen to elicit statistically smaller kidneys that are reminiscent of former times [i.e., before coordinated breeding programs began], the study animals might not be short-changed in the least. If anything, the test animals just happen to more resemble their evolutionary ancestors.) From the HQs that invariably exceed unity, the ERA field next finds

Table 4.3 En route to assembling elements of second-order toxicology.

Perceived biomarker of concern	Reason to monitor biomarker	Specific ERA suspicion or concern	Delta being sought	Study design particulars (to establish the needed delta)
Excess hyaline droplets	Chemically exposed laboratory rodents developed statistically higher hyaline droplet numbers	Affected rodents presumably develop condition x (e.g., abbreviated lifespan)	% increase in hyaline droplets over the norm that causes a (statistically) shorter lifespan	1 Dose rodents to arrive at several distinct classes, in terms of increased hyaline droplet number 2 Monitor animals for lifespan changes
Reduced gonadal tissue weight (gtw)	Fish in nitrogen-contaminated waters have exhibited lesser gtw	Compromised (i.e., lesser) reproduction	Degree of GTW reduction that correlates with lesser productivity	1 Breed fish such that they have lesser GTW; strive to produce fish with a range of % GTW reductions (e.g., 10%, 20%, 40%) 2 Mate fish with reduced GTW to establish the delta
Oversized livers	Oversized livers have been observed in mice exposed to various chemicals through the diet, including a chemical found at a site of interest	None specifically established; it's assumed that 'compromised health' results from the condition	"Default" delta: assume that survival (longevity) and reproduction are compromised	1 Dose laboratory mice with the chemical in question; necropsy animals to verify that oversized livers can be produced 2 Dose a new group of mice, and test mice for development of endpoints of concern

(*continued overleaf*)

Table 4.3 (*continued*)

Perceived biomarker of concern	Reason to monitor biomarker	Specific ERA suspicion or concern	Delta being sought	Study design particulars (to establish the needed delta)
Reduced body weight	Site contaminants are thought to trigger sluggishness and poor feeding habits in birds and mammals	Lesser overall fitness (seemingly)	Degree of (relative) weight loss that compromises a critical biological function	1 Through dietary and activity adjustments, produce reduced weight specimens 2 Observe animals as necessary
Shorter dorsal fin	Site (stream) fish were observed to have the condition	Ineffective locomotion in the water column (presumably)	How much shorter the dorsal fin needs to be before the fish doesn't swim capably	Option 1 Collect affected and unaffected fish; test swimming ability under controlled laboratory conditions Option 2 Breed fish to have shorter dorsal fins; test their swimming relative to controls

itself generating statements about how non-protective contaminated sites may be. Thus, an estimated LOAEL-based HQ of 12 for reduced kidney weight for a fox, marks the fox as potentially at risk. In this scenario however, the LOAEL-based HQ should not have been computed in the first place because it had never been shown that a 15% weight-reduced kidney constitutes an adverse effect (Tannenbaum 2001).

Toxicologists that mean to assist ERA fail us the second time with what would appear to be an inability to introduce improvements to the field and a concomitant demonstrated complacency with a 'more-of-the-same' approach. It is extremely critical to understand this point. Toxicology has indeed supplied changes and even innovations to the way that we assume ERA tasks operationally, but the changes and innovations cannot be described as improvements. The difficulties begin with the standard libraries of toxicity information used in the ERA field (e.g., Sample et al. 1996). By way of example, for many practitioners, if arsenic is deemed to be a contaminant of potential concern (COPC) for a mammal, there is but one choice of a NOAEL and one choice for a LOAEL, each stemming from a 1971 mouse study with arsenite (As^{+3}; Schroeder and Mitchener 1975). For birds, there are two studies that each produced a NOAEL and a LOAEL: one, a 1969 brown-headed cowbird study with copper acetoarsenite (aka Paris Green; USFWS 1969), and the other, a 1964 mallard duck study using sodium arsenite (USFWS 1964). We should marvel at this arrangement. Although there are some saving graces to these studies (e.g., they are genuinely chronic in nature; mortality was the endpoint for each), for the next decades of HQ-based ERA to come, we will likely only be drawing on these same studies should arsenic present itself as a COPC. Wouldn't the most basic and welcomed of improvements to HQ-based ERA take the form of having an expanded series of NOAELs and LOAELs from which to choose – not just for arsenic, but for all the chemicals with which we deal? Consider that if arsenic seems to be a concern at a site, it might not be acetoarsenite or sodium arsenite that is present in the soil, and further that we'll almost never know what form of arsenic we do have at our sites because we so rarely elect to speciate metals. Until such time as HQ method limitations are no longer overlooked, it would be hard to argue against having at our disposal a collection of NOAELs and LOAELS reflecting testing with different chemical forms and a small range of test species. Consider also that new testing would allow opportunities for second-order toxicology, or what we might better term "toxicology's missing link" to finally be brought forward. Thus, aside from knowing that a liver enzyme is under-produced, we would be able to know the upshot of such under-production. Presently, if we know that a certain chemical dose results in a 15% under-production of an enzyme, we don't know that this is a deleterious effect.

Why ERA practitioners do not decry the slim pickings we have regarding NOAEL and LOAEL availability is puzzling. There may be a feeling amongst practitioners that knowing that everyone is using the same study more than compensates for the reality that for a given chemical there is only one study (and therefore one NOAEL or LOAEL) to select. Alternatively, confidence with repeatedly using the same studies might be augmented when we see the 'slim pickings studies' considered in a broader context, such as the benchmark dose (BMD; US EPA 1995; US ACHPPM 2000). Here, toxicologists supporting ERA comb the literature for studies that meet their rigorous selection criteria as studies of technical merit, focusing on only certain key toxicity endpoints. Though the studies span many species (e.g., for mammals: rodents, rabbits, cats, dogs, monkeys) and variable forms of a given chemical, plotting the effect level doses (effect dose, ED; effect concentration, EC) for a particular endpoint, may with some data smoothing, arrive at a common toxicological expression, such as an ED20 (effect dose for 20% of an exposed population).

For all of its novelty, the introduction and application of the BMD approach to ERA in recent years is a great example of toxicology failing to break free of its ways, and to therefore sell us short in our efforts to assess health risk potential. What can one *effectively* (pardon the pun) do with an EC20 or any other ED? Using a food chain model, one could estimate a receptor of concern's intake of a given chemical and find the intake to exceed a given ED. Where would this leave us, though? Such a finding would not empower anyone to say with any confidence at all, that a site receptor is at risk, or is experiencing a toxicological effect of concern. Such ED exercises only serve to put into broader context the 'more of the same' contributions of toxicologists to the field. Thus, those working in toxicology as it assists ERA only seem capable of contributing to the ERA process component of effects *screening*. We can only speculate as to why toxicological know-how hasn't been put to better use, enabling us to encroach on the question of whether or not ecological receptors at contaminated sites actually present with health impacts. It could be that the field as a whole is unmindful that screening outcomes are not demonstrations of ecological problems. Screening outcomes couldn't possibly be reliable demonstrations of health effects when they review chemicals one-by-one, extrapolate from one-generation studies to the field condition (of tens and hundreds of generations already having been exposed), and deviate from the field condition in other ways that have been discussed earlier. It could be that more aggressively tackling the ecological risk question is perceived to be too insurmountable a task. Regrettably (for ERA, and for science overall), it might be that screening for effects has been accepted as the goal of ERA, although to our knowledge no regulatory agency has ever openly espoused such a notion. The situation is regrettable because screening alone leaves us ill-prepared for remedial decision-making. It is not a challenge to conjure up possible bases for screening assessment outcomes being viewed or understood to be goals or endpoints in themselves. It is clear that sites do proceed to the point of invoking cleanup actions or instituting ongoing monitoring programs where only screening efforts have proceeded. In accounting for such practice, we may suggest the following. Regulators are uncomfortable with the notion of contamination being left in place. Due to strong-arming or other means, the process has clearly been worked to a point where it is more or less a given that screening outcomes are accepted as laying a sufficient enough basis for invoking remedial measures and the like – all to the delights of the regulating community. Further, where resistance to acting on the outcomes of crude screening efforts might be expressed, regulators, in defense, can easily point to any number of sites where screening outcomes were all that were needed to assume the worst ecologically.

We would do well to consider how deeply entrenched in the screening mode ERA remains where the science continues to be non-attuned to knowing with reasonable confidence that actual deleterious toxic effects are taking hold in the field. There is perhaps no better example of the entrenchment than the US EPA's ecological soil screening levels (Eco-SSLs) initiative (US EPA 2003c), a collaborative effort of a multi-stakeholder workgroup consisting of federal, state, consulting, industry, and academic participants led by the EPA's Office of Emergency and Remedial Response. The initiative's intent has been stated various ways:

- to help risk assessors focus their resources on key site-specific studies needed for critical decision-making;
- to decrease the possibility that potential risks from soil contamination to ecological receptors will be overlooked;
- to increase consistency among screening risk analyses;
- to avoid underestimating risk.

En route to achieving its intent, for each of four distinct ecological receptor groups (plants, soil invertebrates, mammals, avians), a four-step process is applied. First, a comprehensive literature search in undertaken. This is followed by screening the literature with exclusion and acceptability criteria to cull only those studies that have high scientific merit. Applicability of selected studies for submitting to Eco-SSL derivation, the third step, is determined through extracting, evaluating, and scoring test results. The fourth step is the actual derivation. The overall effort may seem impressive and there may be a nice ring to the term "Eco-SSLs" that is now part of our lexicon, but Eco-SSLs reflect no more than the HQ methodology replayed. As the guidance (US EPA 2003c) states, wildlife Eco-SSLs are the result of back-calculations from a HQ of 1.0. The guidance also provides an accepted and generally true statement – that a HQ of 1.0 is the condition where the exposure and the dose associated with no adverse chronic effects are equal, indicating adverse effects at or below this soil concentration are unlikely. Is this not something that we already know? It's not clear what we should marvel at first with regard to the Eco-SSL tool. Is it that toxicologists supporting ERA have put so much of their time into such a non-sophisticated effort, that is, alas once again, only a screening one? Is it that we have gained no ground for that most likely case of a soil concentration exceeding an Eco-SSL, informing that the concentration corresponds to a HQ greater than 1.0? For that particular case, we still don't know if a HQ greater than 1.0 means that receptors are presenting with adverse effects. Perhaps then, we should marvel that we've gained no ground in our toxicological understanding of what may be occurring at contaminated terrestrial sites.

We should also marvel at the Eco-SSL effort for several other reasons. At the time of this writing, it is more than a dozen years since the effort got underway. Since the intended list of chemicals to submit to the Eco-SSL initiative was never more than 24 (as 17 metals and 7 organic contaminants), and with the list having been gradually compiled over several years, we are prompted to ask what has taken so very long to furnish the numbers. With an embarrassingly slow track record, we are prompted to ask if Eco-SSLs are truly needed at all. We might consider how many ERAs have been completed since the year 2000, when the effort got underway. To our knowledge no ERAs have ever stated in their respective uncertainty sections, that an assessment shortcoming was the inability to conduct an Eco-SSL screening for the any of the profiled 24 chemicals that were present at their subject sites. We might also consider the exceedingly high frequency with which ERAs have produced HQs greater than 1.0 over the years. If a chemical was found to have a failing HQ for a specific receptor (seemingly reflecting the condition for other animals in its feeding guild), we see retrospectively, that the non-availability of Eco-SSLs at the time an ERA was being conducted, didn't hamper the assessment process in any way. To the contrary, screening with Eco-SSLs, were they to have been available, would have constituted an unnecessary step. With Eco-SSLs intended to be conservative values, we would absolutely marvel at the unlikelihood of a site of concern having all of its contaminants screen out with Eco-SSL application. The reader is reminded that running HQ calculations with the lowest concentrations of inorganics known to exist in US soils, and considering only that portion of the diet that is incidental soil ingestion (approximately 10% for birds, 2% for mammals), produces failing HQs (Tannenbaum et al. 2003). What emerges from this discussion is that the still ongoing Eco-SSL initiative is a self-fulfilling prophecy of sorts. To the extent that regulators want to see that all sites at least initially have a problem so that they can submit to lengthier evaluations, toxicology (in deriving Eco-SSLs) has allowed for the design of a process that accomplishes just that. Application of Eco-SSLs ensures that all sites will submit to the formal ERA process, one that just happens to be HQ-based.

There is an easy way to gain some traction on this topic of hoping to see toxicological know-how applied in a considerably more constructive fashion for sites that concern us. Gaining traction involves no more than acknowledging that any given site to be assessed from this point onward (effectively all sites that will ever submit to ERA) will present with one or more HQs greater than 1.0. It's a fair bet that such will be the case. Although it isn't true, let us even say that the screening to be conducted, taking either the form of comparing site media concentrations against established effects-based benchmarks, or calculating HQs, is credible and highly informing. As has been stated times before in this and other chapters, the skill sets of toxicologists are seemingly needed, for we cannot advance to remedial decision-making based on the outcome of any *screening* work. With all of the attendant uncertainties built into the HQ computation, we don't know enough to 'convict' sites, i.e., to say that cleanups are needed because significant toxic effects are arising, or have already arisen in site receptors. To the "Now what?" question posed to toxicologists (i.e., What do we do now that we very loosely suspect that site receptors are experiencing undue stress?), we will likely hear back that HQs can be refined by changing exposure assumptions in the computations. Indeed HQ estimates can be refined, but in proceeding this way we are once again only screening. But ERA is not only about screening; unless we've lost sight of the goal (and we may have), we're out to know what the actual risk levels are for site receptors developing certain toxicological endpoints. We will never encroach on this, however, if any and all of our efforts are linked in some way to the HQ computation. To get down to brass tacks, we must consider whether or not our toxicologists are capable of breaking free of HQ thinking. It is only after making that break that we can ever look forward to the development of novel approaches that can lead us to the information we need.

It is interesting to note that in the case of aquatic receptors, we *do* advance beyond screening. If benchmarks for survival or for non-lethal health effects (e.g., deformities, fin erosion, tumors) are exceeded, we know to advance to the field to see if indeed fish are present in reduced numbers, or that fish present with deformities, etc. Certainly for the terrestrial investigations that consume us, we are not so oriented. Perhaps we should be baffled by this. We can, of course, construct a list of reasons for being disinclined to advance to the field for a look-see, and heading that list would probably be that it's easier to catch fish than it is to catch birds or mammals. Legitimately, animal care and use requirements would pose their share of complications too, but there are means for dealing with these. The rank and file of toxicologists supporting ERA has apparently opted to not pursue a field-based assessment scheme for mammals paralleling what is done for fish and other aquatic species, and this is our loss.[7] The reader

[7] The reader is reminded that Chapters 8 and 9 discuss field-based ecological assessment.

should not think that instilling advanced toxicology in ERA is as simple a matter as collecting animals from the field. Should we ever break out of the desktop HQ-based approach to assessment that invariably leads to a dead-end, and elect to cull specimens from the field, the challenge to the toxicologist will be to render a health status determination. This must be clear; the goal would not simply be to pronounce that site receptors are (statistically) different from their reference location counterparts at some level of biological organization (cell, tissue, organ, biological system or function).

An ERA investigation at a military installation from ca. 2004 will serve to consolidate ERA's toxicology bottleneck. Among the site's numerous soil contaminants was perchlorate that had only a few years prior burst upon the chemical health risk assessment stage as a formidable contaminant of concern. Perchlorate was regularly turning up at military installations because it had been used for decades as an oxidizer in solid rocket fuel and munitions. At the time, perchlorate was without an available avian NOAEL and LOAEL, and the federal and state regulators involved with the site project were uncomfortable moving onto remedial decision-making with such a toxicological data gap on their hands (i.e., the inability to generate HQs for the chemical). Just at this time, the contractor assigned to the project happened by a perchlorate paper that had just been published (McNabb et al. 2004). The paper was heralded as a great treasure by the project team and particularly by the regulators, for it would now allow for HQ computation. Perchlorate HQs were hurriedly computed for growth as the toxicological endpoint, and the avian HQs exceeded 1.0 for a good many soil sampling locations. According to the HQ analysis, birds were at risk of compromised growth from their onsite exposures. Or were they? In the study, ammonium perchlorate (AP) had been administered at two discrete doses in drinking water to 3- to 4-day posthatch bobwhite (*Colinus viginianus*) for either 2 or 8 weeks. The study first evaluated organismal thyroid status (circulating thyroid hormones) because perchlorate is a known thyroidal iodide uptake inhibitor in vertebrates (Taurog 1996; Wolff 1998). Disrupted thyroid function/thyroid hormone deficiency can be linked to growth effects, and thus skeletal growth as indicated by femur and tibia length was additionally monitored. The ERA team seized on the femur and tibia length measures of the 8-week exposure study arm, and not on reduced plasma thyroxine concentrations, increased thyroid weights, or reduced plasma thyroxine content of the paired thyroid glands, all of which were far more demonstrably compromised. Presumably this choice reflected growth being an established toxicological endpoint of concern in ERA. A LOAEL for decreased tibia length was established for the 8-week study to suit the needs of the installation's ERA (US ACHPPM 2007) but curiously, the basis for doing so did not correspond to the study authors' reckoning of the experimental outcome that is provided here.

> Our measurements of thyroid-responsive endpoints (body weight and limb growth) suggest these indices are not altered until the tissues have had sustained exposure to decreases in thyroid hormones. Thus these indices are very insensitive ones that have little practical application for assessing alterations in thyroid function. Even in growing bobwhite chicks exposed to high AP for 8 weeks, body growth was not systematically decreased by the AP concentrations used (up to 4000 mg/L). Femur and tibia growth, which are sensitive to thyroid hormone decreases in chicken embryos exposed to polychlorinated biphenyls, were not altered compared to controls except at very high AP concentrations ($\geq$2000 mg/L for femur; 4000 mg/L for tibia) after 8 weeks of exposure.

As an overarching observation (one it is hoped that the reader has made), we see in this true account of the site remedial process, that HQ computation is really all that a project team cares about. We see this in the hesitancy to move the project forward without HQ computation, and in the members of the project team being unaware that the perchlorate-caused 'growth' effect (assumed to have taken hold given the HQs >1) amounted to a lesser tibia length measured in the very young birds. In the quest to produce HQs, we see that liberal interpretation of toxicological information taken from *bona fide* studies is not only possible but perhaps likely to occur. The project's contractor, pleased to be able to derive the LOAEL that the regulators felt was so sorely needed, defended his derivation in the following way (paraphrased here): "Reduced bone growth (here, the tibia) and delayed bone ossification are reflective of body growth. Since body growth (then) was indirectly arrested, and with body growth being an established toxicological endpoint of concern in ERA, we have the basis for bringing the derived LOAEL, and thereby a new TRV, online."

Before adopting the new avian LOAEL-based TRV, we should ask if a lesser tibia length translates into a smaller bird when development ceases and a bird attains its full-grown status. We should ask this because upon hearing that growth was reduced, most would expect this to be referring to an animal's overall size or height. We should also ask if tibia growth reduction is deleterious; is there some biological function that is compromised when a tibia is undersized? Can a bird with a shorter tibia run along the ground properly, take to flight, or alight on a branch or fence-post as it should? To be fair, the mechanics of movement, posturing, and flight could be compromised with a shorter tibia (and presumably two of them; one in each leg), but reasonably too, the change to this bone might actually be advantageous; a shorter tibia should mean a lesser body weight, something that is most conducive to sustained flight. We might ask how much shorter a tibia needs be before the 'defect' is recognizable to a conspecific, such that a conspecific would relate differently than it should to its tibia-challenged companion. We should not be surprised to learn that the McNabb (2004) study did not extend to maintaining the perchlorate-dosed bobwhites until they attained their full maturity. We also don't have the luxury of knowing if the reduced tibia effect disappears if perchlorate, as a stressor, is removed during the young life of a bird. In the McNabb 2004 study, it was only the highest test dose (of 4000 mg/l), supplied daily for 8 weeks, that brought on the effect. Realistically, a bird at a perchlorate-contaminated site might not continuously drink from the same contaminated ponded seep. Further, even if a bird should only utilize one contaminated drinking source, the perchlorate concentration there might not be so very high. In a context of the many questions raised here regarding the LOAEL that was readily developed and applied in the site-specific instance, the reader should note that the observed reduction in tibia length after 8 weeks of high-dose exposure was all of a statistically significant 2 mm relative to the study controls. The reduction was real and undoubtedly reproducible, and to our understanding, perchlorate in the drinking water was the causal agent. What though can we say of the site's remedial project team that stood by the HQs >1? Earlier we were unimpressed with the team not knowing that the growth effect was a shortened tibia. Now that we see how very minimal of an effect, or more correctly a change, there actually was, we probably should be unimpressed too that the 2 mm difference figured as prominently as it did in the remedial decision-making.

The above case study comes to illustrate how more often than not we apply the toxicology information that we have at our disposal. Here in the eagerness to apply a TRV (one that became available when it was thought to be most needed), the parties that should have known the basis of the toxicological effect being selected for, were found to be completely uninformed. Lost sight of, is the fact that only a potential-for-risk screen was conducted. The information to make a defensible case for birds being at risk from perchlorate exposure had not been assembled, and there was no effort afoot that we know of, to allow for such a thing. Stepping back from this profiled case, we will find the larger problem of our misjudging our capabilities to assess risk. As stated earlier, an HQ in any form is only a screen, and not a measure of risk (US EPA 1989; Kolluru 1996). This point continually fails to resonate within the toxicology and ERA camps. And so, every toxicologist supporting ERA who believes that critical gaps in our assessments can be filled by conducting the right dosing study or by compiling the dose–response information of an expanded set of previously conducted studies, is our greatest enemy. The lacking sense that HQs, in any form, are not risk expressions is never more apparent than in an excerpt of a recent published article that sets forth a distinction between what it terms "a hazard (or screening) assessment" and "a risk assessment" (Allard et al. 2009).

> A hazard assessment typically compares point estimates of exposure with a toxicity metric that is considered likely not to result in an adverse effect (i.e.,

> the TRV). This is to provide decision-makers with a single point estimate as a HQ. Interpretation is based on whether exposure is above, at, or below the TRV (i.e., whether the HQ is above, at, or below 1). Given the potential for uncertainty in the estimation of the exposure and toxicity metrics, liberal interpretation is used when quotients are close to a value of 1. Conversely, a risk assessment combines distributions to provide decision-makers with information about the magnitude and probability of a range of outcomes. Interpretation is based on differing probabilities of an adverse outcome. What distinguishes a screening assessment from a risk assessment is that the former does not explicitly and quantitatively address the probability of an outcome, while a risk assessment does.

For all of the article's well-intended definitions, screening level ecological risk assessments (SLERAs) and baseline ecological risk assessments (BERAs) are not probability statements, and they never can be. They can tell us crudely how much more than a safe quantity of a chemical an animal is taking into its body, but the simple ratios cannot provide an indication of how *likely* it is that an animal will present with a health condition. The infamous "full blown ERA" bears mention here. Curiously this terminology, to the author's knowledge, is not captured anywhere in ERA guidance. It is though, accorded the highest esteem as in the parlance so often heard in negotiations: "Well, maybe we need to have this site move on to a full-blown ERA". And threats can be made with this curious term as well, as in: "If we can't come to some agreement, we're going to have to move onto a full-blown ERA, and I was hoping we wouldn't have to go there." There is no such thing as a "full-blown ERA", as in some all-encompassing treatise that unequivocally informs of the probability that a receptor is experiencing risk. What people are terming "full-blown" ERAs are documents that terminate once again in HQs, albeit ones that are more refined and seemingly more truthful than those of screening-level efforts. BERAs do not provide probabilities of receptors developing health effects; they might provide probabilities of arriving at certain magnitude HQs.

With its overextended loyalty to the HQ computation, toxicology has failed us. As the previous case spotlights, toxicology has also failed us in a second, more basic way, with a glossary of inconsistently defined and applied terms. In the earlier tibia length debacle, we must throw into question if it was a Lowest Observed Effect Level and not a LOAEL that was being considered. More shockingly, the LOAEL that was used to assist the project team (and that unfortunately stands to be used in other avian assessments; US ACHPPM 2007) might actually have been a NOAEL. More empirically, it was not a biologically adverse effect that formed the basis of the decision making, but rather a statistically significant one. Readers would do well to engage in the relevant literature that perhaps best clarifies the terminology that is so central to toxicologically based ERA work (Lewis et al.

2002), and to then review the studies that form the basis of TRVs they elect to use. This should amount to an eye-opening experience.

We should appreciate the area of high interest among researchers in the environmental field in pointing at or assuming linkages of body burdens and toxicological effects. We are referring here to a vast and growing literature that conforms to a now easily recognizable template. The reporting takes the form of measuring a chemical in some biological tissue, reporting an observed trend with regard to such chemical detections (perhaps with chronological time, or with increasing distance from a contamination source), and suggesting that the accumulations could well signify health consequences in the future. Table 4.4 facilitates a brief discussion here. For simplicity's sake, the table was assembled using publications drawn from the same journal and from a brief span of years. All are valuable contributions to the peer-reviewed literature and to the global body of knowledge of contaminant accumulations in a broad array of species. Probably none of the scientific papers of this genre, however, relate to Superfund-type risk assessments in a hands-on way. Vis-à-vis our needs, the subject matter of the papers does provide a certain academic stimulation that ERA could do without. Reports of chemicals appearing in certain tissues and reports of mounting concentrations developing with advancing time all come to suggest that there is great purpose to our ERA efforts. In other words, because contaminants are capable of being taken up, or actually do bioaccumulate, we must be unrelenting in our pursuit of possible severe ecological impacts occurring at hazardous waste sites and the like. As we have seen, our sites are frighteningly small and thereby, to begin with, do not regularly support a great many of the receptors that truly concern us. Since we won't be measuring body burden in site receptors, and since we have no way to relate body burden in ecological receptors to health effects anyway, we should not be letting these studies, that encompass vastly larger geographical areas, impact on the limited science that is ERA's. Just because contamination is present – in inanimate environmental media or in living tissue – we don't know that animals are at risk and we shouldn't necessarily assume that they are. A perusal of the "conclusions" of Table 4.4 doesn't indicate that we've uncovered anything more than the anticipation that health effects might eventually set in. Ecological assessment for the run-of-the-mill 5-, 10-, or 20-acre contaminated sites with which we deal can't be governed or held hostage by the precautionary principle (Appell 2001; Marchant 2003; Tickner et al. 1998).

As a next-to-last consideration for this chapter, we will consider the toxicologist as a reviewer of data, and the potential to err when acting in this capacity. The parent material for this discussion comes from the author's research in the area of reproductive toxicity in rodents. The backdrop to the research is the recognition that numerous components of the female reproductive system can be offset from chemical exposure. Consistent with the earlier expanded discussion on biologically

Table 4.4 "Guilt-by-association" – reporting contaminant uptake in biological matrices, even if intended only to be descriptive, spurs on ERA at Superfund-type sites.

Chemical exposure issue	Essential finding*	Conclusion†	Reference
Accumulations of perfluoroalkyl contaminants (PFCS) in East Greenland polar bears	Increasing liver and kidney accumulations	If trends continue, rat and monkey NOAELS and LOAELS will be exceeded	Dietz et al., 2008
p,p′-DDE and PCB in Bluefin tuna	Significant tissue accumulations over trophic levels	Significantly higher ontogenetic magnification factor now known	Corsolini et al., 2007
Organochlorines and brominated flame retardants in tawny owl eggs	Significant body burden declines	Body burden reductions result from reduction in retardants usage	Bustnes et al., 2007
Polybrominated diphenyl ethers (PBDE) in red fox	Higher accumulations in fox than in voles and mice	BDE 209 unambiguously bioaccumulates in terrestrial top predators	Voorspoels et al., 2006
2,3-dibromopropyl-2,4,6-tribromo-phenyl ether, in seal blubber	The compounds penetrate the blood–brain barrirer	Compound biotransformation now understood	Von der Recke and Vetter, 2007
Arsenic in bark beetles	Woodpeckers accumulate arsenic from feeding on contaminated bark beetles	Significant accumulation and transfer in food chain may threaten birds	Morrissey et al., 2007
Trichloroethylene (TCE) in apple and peach trees	TCE only detected in roots	Potential TCE transfer to leaves and fruit needs more study	Chard et al., 2006
Polyaromatic hydrocarbon (PAH) and PAH metabolites in oiled guillemots	PAH composition in liver corresponds to number of benzene rings in compounds	Evidence of presence and stereoselectivity of hepatic microsomal CYP1A1 in guillemots	Troisi et al., 2006

*Finding or anticipated outcome.
†Conclusion or supposition.

significant thresholds (as opposed to statistically significant ones), the objective was to identify that degree to which a component of the reproductive system (ovarian follicle count) would need to be offset to trigger impacts. Briefly, female mice were dosed with a pesticide known to actively target developing follicles (i.e., those that would otherwise support fertilizations taking place in the future). Indeed, a mating study involving (female) mice with an "engineered" reduced ovarian follicle count (produced through dosing with the pesticide) yielded statistically significant differences (Table 4.5). How would/should the Table 4.5 results be interpreted? We could probably expect the toxicologist working in ERA to conclude that the low- and high-dose testing both support LOAEL development for reduced (paired) ovary weight, with the accompanying thinking being that such reduced ovary weights are harbingers of compromised reproduction; realistically, the lesser ovary weights in the two test groups were not due to chance. Of course, a required lesser ovary weight to trigger lesser reproductive output (e.g., fewer or smaller-sized litters) has never been established heretofore. Presumably at some point, the weight is too low and reproduction *is* compromised. Let us now review Table 4.6, focusing on its additional column of data. We find that litter size (i.e., young production), a purer reflection of reproductive success, was not statistically different between the test groups and the controls. What we also find is that had only Table 4.5 been provided, there are those who would have derived TRVs for ovarian follicle count reduction as an endpoint, and understood it to be a detrimental effect of concern – when it is not. The described case alerts us to the trigger-happy nature of toxicologists to latch onto toxicity study data and churn out unfounded TRVs and the like. We might label the Table 4.5/Table 4.6 sequence "Demonstrated toxicity: now you see it; now you don't."

This chapter's review of toxicology as it pertains to ERA would be far from complete if it did not broach an all-encompassing question that evidently eludes ERA practitioners: Are contaminated media toxic to ecological receptors at all? From the author's perspective, there is more than enough reason to believe that

Table 4.5 Summary of ovary weights and litter data by group.

Pesticide dose group	Ovary weight (g) (paired; mean ± SD)	Total number of dead pups	Total number of resorptions (early/late)	Individual pup weight (g) (mean litter)
Control	0.0429 ± 0.0141	0	0/0	1.174 ± 0.199
48 mg/kg/d	0.0295 ± 0.0064*	1	1/1	1.130 ± 0.213
96 mg/kg/d	0.0312 ± 0.0036*	0	0/1	1.012 ± 0.217

*Statistically significant difference from controls (one-way ANOVA on ranks, $P > 0.05$).

Table 4.6 Summary of Ovary Weights and Litter Data by Group.

Pesticide Dose Group	Ovary weight (g) (paired; mean ± S.D.)	Total Number of Dead Pups	Total Number of Resorptions (Early/Late)	Indivuidual Pup Weight (g) (mean Litter)	Litter Size (Live Pups; mean ± S.D.)
Control	0.0429 ± 0.0141	0	0/0	1.174 ± 0.199	11.222 ± 2.949
48 mg/kg/d	0.0295 ± 0.0064*	1	1/1	1.130 ± 0.213	10.200 ± 2.530
96 mg/kg/d	0.0312 ± 0.0036*	0	0/1	1.012 ± 0.217	9.556 ± 2.297

*Statistically significant difference from Controls (one-way ANOVA on Ranks, $P > 0.05$)

contaminants sequestered in media exude toxicities that are considerably attenuated relative to what we would otherwise think. Bringing such considerations to an extreme and where it could be demonstrated that chemical toxicity is indeed markedly reduced, we might discover that chemical exposure risk assessment for ecological receptors is never needed. With this alternative understanding, we could gracefully exit the ecological assessment field instead of barraging people to cease with applications of the established ERA paradigm through repeatedly exposing the many weaknesses that plague it. In short, we should be open to finding that the essential ERA question (Are site ecological receptors in danger of experiencing toxic effects from the chemicals they encounter?) doesn't even start.

A quest to demonstrate that contaminated media in the field are not necessarily toxic to the biota that contact them is supported by the recurrent case whereby sites do not seem to have a challenged or perturbed ecology. A chemical fate and transport consideration lends additional support. With sites being 30 or more years old, there has been enormous opportunity for contaminants to have substantially weathered or entirely disappeared.[8] That studies have been conducted to demonstrate reduced chemical toxicity with time, alluding to the phenomenon of risk overestimation, is somewhat of an understatement (Alexander 2000). For all of the interest they have generated and the information they have furnished though, the limitations of the studies show through. Studies designed to investigate and report on the modified toxicity of aged chemically amended soils have shared two repeating features. First, they have largely defined or demonstrated reduced toxicity through the tracking of diminished (i.e., time-dependent) chemical bioavailability (Alexander 2000), as for example in the analysis of tissue or body burden (Alexander 1995; Tang et al. 1999; Umbreight et al. 1986). Second, because the studies have by and large been simulations, the various designs have reflected substantial departures from the ambient condition

[8] This includes the case of organic compounds, such as pesticides, that are relatively short-lived in the environment because they are subject to photodegradation.

in a host of ways. Reaction containers have ranged from test tubes to small plant pots, and testing has proceeded after only days or weeks from the time of chemical addition to soil. Nearly all studies have taken place indoors (amidst fixed temperature and artificial lighting regimes), and acute rather than chronic toxicological responses have been monitored. Finally, test organisms have often included species that are not of concern in the health risk assessment arena (e.g., bacteria, fruit fly, springtail, enchytraied worm).

In contrast to the above, the author's research has study design features that vastly improve in approximating the real-world condition: (1) large reaction containers (kiddie pools measuring aproximately 6 ft in diameter and 2 ft high) holding hundreds of pounds of soil at a time, far surpassed any relevance that might be associated with test tube use; (2) reaction containers were not shielded in the least but rather exposed to a fully unimpeded outdoor/ambient condition; and (3) the outdoor exposure period extended to a full calendar year. Two chemicals were tested, lead acetate and the 2,4-dinitrotoluene (DNT), and as the reader will later learn, the basis for their selection impacts greatly on the study's overall findings. Two radically different soil types widened the experiment test matrix (and, as it turned out, also strengthened the confidence in the findings). For now, the reader should understand that budgetary constraints meant that only two chemicals could be tested, one metal and one organic species. Lead was chosen due to its ubiquitous nature as an environmental contaminant arising from multiple release sources (e.g., battery disposal, smelting, manufacturing, oil refining, etc.). Lead acetate was specifically chosen because it most often forms the basis of the lead-based TRVs for birds and mammals (Sample et al. 1996). The choice of 2,4-DNT reflected the interest at the time, to evaluate a military-unique compound. With the (legal and other) complexities associated with acquiring appreciable quantities of an explosive needed for amending soils in the kiddie pools as reaction containers, the specific choice of 2,4-DNT was driven only by its relative accessibility by the contractor conducting the work.

Procedurally, range-finding tests to produce effects concentrations (as EC25s and EC75s) in four species (the earthworm, *Eisenia fetida*; lettuce, *Latuca sativa*; barley, *Hordeum vulgare*; Northern wheatgrass, *Elymus lanceolatus*) and for two widely divergent soil types were conducted in the laboratory. The EC concentrations were then established in the large soil volumes of the kiddie pools, followed by setting the soil-amended pools out in the field along with other pools of non-amended soil (to serve as controls, etc.). After the year-long exposure, the soils were brought indoors, amended soil concentrations were determined, for chemical loss that might have occurred, attributable to chemical breakdown and/or percolation/drainage through the soil column. For lead, the post-weathering concentrations for the two soil types were 75% and 96.5% of the initial concentrations. For DNT, the respective soil types lost 92.8% and 97% of their initial amendments (a not unexpected finding). The post-weathering concentrations were then established in pools of previously non-amended soil that had weathered outdoors. At this point, standard toxicity tests involving as many as seven endpoints (for plants, the endpoints were emergence, shoot length, root length, shoot dry weight, and root dry weight; for earthworms, the endpoints were survival and reproduction) were conducted. The essential comparisons were of the toxicity responses of control soil with (a) soil that had weathered with the ECs, and (b) clean soil that weathered and that had been subsequently amended with the ECs. Regarding lead, cumulatively for the plant and earthworm testing (i.e., considering all EC25 and EC75 concentrations for all endpoints and for both soil types), 80% of the (240) comparisons found no statistical difference between chemically amended and control soils. Regarding the selected explosive, plants exposed to weathered 2,4-DNT-amended soil were indistinguishable from those in control soil 98% of the time. The lead findings are particularly telling since this chemical did not disappear. How is it that lead that surely produced toxic effects in the range-finding tests and that was present at concentrations as high as 11,200 mg/kg in amended soil, did not shortchange plant emergence and a battery of growth and development endpoints in the least, in multiple species and for distinctly different soil types? The study again, was not designed to identify the fate and transport feature(s) that might account for, in this case, a vastly reduced toxicologically elicited response. We might surmise that the lead binds tightly and expeditiously to the soil moiety such that it cannot be bioavailable (with bioavailable here used in the more traditional sense, and not as the term is applied in Chapter 3). At the very least, we discover that just because a chemical is present in soil, and even where the concentration is highly pronounced, toxicity need not necessarily be linked to it. There is more to glean from the 2,4-DNT component of the study than merely reporting complete loss of the compound after the weathering year, and the concomitant cases of no statistical difference in control soil and weathered soil toxicity test comparisons. Since the pesticide had all but disappeared, the few cases of (significant) lesser response must mean that it was not the amended DNT that impacted growth, etc., but quite obviously a breakdown product of DNT. We learn then, that there are cases where we can (and do) ascribe an effect – and derive a toxicity factor – to a given chemical that we administered, although the effect is not (i.e., cannot be) actually attributable to it.

These results strongly suggest that repeatedly in health risk assessment (that is, in both HHRA and ERA) critical mismatches of toxicity factors and site-specific instances of contaminated media are occurring, as follows. Invariably, chemicals used in the dosing studies from which toxicity factors will ultimately be derived (TRVs for ERA; cancer slope factors and reference doses in HHRA), are brand new. In the jars in which they arrive from the commercial supplier, the test chemicals have a maximum potency or toxicity born of the purity of the produced lot (which is indicated in the packaging label) and owing to the unreacted nature of the chemical (i.e., not even having been exposed to the open air). When first mixed into the diet or drinking water, topically applied, or amended to soil, test chemicals have their best chance of eliciting effects. (In the described study, the EC25s and EC75s of the short-term, laboratory-based range-finding tests, reflect this initial heightened potency.) What of the chemicals though, in the soils of Superfund-type sites? These chemicals are anything but new and purified. For 30 or more years they have been exposed to temperature and humidity extremes through the cycle of seasons, had precipitation come upon them, and had ionizing radiation sear them, all the while as part and parcel of the soil matrix. Additional opportunities for chemicals to bind with soil particles will have been afforded through soil compaction events, as in soils having been walked or driven across. Finally, soil-bound chemicals in having passed through the gut of earthworms or other receptors, will have had still more opportunity to have had their toxicity altered and effectively reduced. When a conventionally derived toxicity factor is applied to the case of a receptor exposed to a dated contaminated site, we are assuming that the decades-old and highly weathered chemical is still eliciting its maximum toxic punch? Why? Why, when there is every reason to expect – and now *know* by virtue of the upscaled chemical aging study described above – that the chemicals in the media of concern are so toxicologically removed from those used when dosing animals for the purpose of securing toxicity factors?

The purpose of the chemical aging study was not to identify the specific physico-chemical soil factors or the mechanism responsible for the drastically attenuated if not completely absent toxicity observed. With the suspicion that aged chemicals become dramatically less toxic now validated, we see that toxicology, intent on assisting ERA, is in reality leading us further astray. We should ask why toxicologists haven't addressed the critical mismatch. Admittedly, adjusting derived toxicity factors to reflect the real-world, aged-chemical condition of Superfund-type investigations is no simple matter. More than likely though, toxicologists haven't addressed the mismatch because they never considered that such a phenomenon is so operative. With each new toxicity study conducted for the purpose of furnishing a TRV, where reduced toxicity with time is a concept not incorporated, we see that the toxicologist is not attuned to ERA's needs.

Perhaps for the skeptical reader, the results of the profiled study are not so convincing. After all, only two chemicals were reviewed. Herein though, is an opportunity to see that with the minimal information we have culled from the study, we need not be so distrusting of the discovered phenomenon that appears to be so utilitarian to our needs. For the open-minded reader, there is 'Angela Shelton science' or 'Angela Shelton ecology' to consider, as follows. Angela Shelton is a multi-talented individual; a gifted actress, comedienne, and film/documentary maker. Several years ago, her creative energies led her to do something quite out of the ordinary. For a documentary she hoped would eventually fall together, she sought out other women who shared her name, and did the leg-work to meet with and interview them. She initially contacted 76 Angela Sheltons, 40 of which, in 40 different US states, agreed to meet with her. Shockingly, there was a commonality to the bios she collected. Twenty-four of her namesakes had been abused (victims of attempted or actual rape, child sexual abuse, or domestic violence). An additional four of the 40 Angela Sheltons interviewed came forward to report that they too had been so abused after the documentary ("Searching for Angela Shelton") had been completed. In all, 28 of 40 interviewed Angela's (or 70%) had suffered the tragic inhumanities, a figure that shatters what is commonly held by experts working in this area.

How do the Angela Shelton statistics relate to the finding that lead and DNT in soil have vastly reduced toxicity after aging for but a fraction of the time that we know chemicals to exist at Superfund-type sites? The answer lies in recognizing that there is statistical power associated with purely rare and random events. We would be more than hard-pressed to say that the 70% statistic is purely a fluke, i.e., that it is just by chance for the name 'Angela Shelton', that such horrific statistics apply. Of course, most of us have no more than toyed with the notion of meeting others that share our names altogether, and perhaps a few of us have at one time or another actually met up with an unrelated namesake. Further, it is most unlikely that another imaginative filmmaker would invest the time and other resources to repeat what the filmmaker Angela Shelton did here, since the idea would no longer be an original one. There would be bias too in setting out to specifically ask individuals if they had been victims of sexual abuse, when that wasn't the premise for the film that was completed; the disturbing accounts of abuse came as a complete surprise to the Angela Shelton filmmaker. We should trace down the purely random nature of the documentary's finding. First, it would be unreasonable to suggest that having the name Angela Shelton predisposes one to become a filmmaker. It is similarly unreasonable to suggest that only one having this name would culture the notion of meeting up with one's namesakes, and actively set out to do just that. Finally, we won't believe that having the name Angela Shelton predisposes a person to be a victim of the abuses discussed here. In other words, we won't believe that had the one female artist to assemble the film had any other name, the percentage of those interviewed would be some 20% only. Hence, the frightening 70% victim-of-abuse statistic is more real than we might wish to acknowledge, but more importantly, highly descriptive of the horrific behavior under discussion here.

With regard to the chemical aging study, it would be a forced suggestion to say that it is only for lead (as a metal) and DNT (as an organic species), that vast toxicity reductions occur in year-old soils, i.e., that by some fluke, the two chemicals that were used in the chemical aging study just happened to produce the dramatic findings that would appear to have key implications for ERA. Lead was selected from a pool of more than 20 other metals, and for a most legitimate reason as was mentioned earlier. The list of organic species (that includes pesticides, solvents, and PCBs, in addition to explosives) runs considerably longer than this, and DNT ended up being used in the study only out of convenience. Reasonably, had other metals or organic compounds been similarly tested, the findings should be the same as those for lead and DNT, and to the point that there is no legitimate need to repeat the testing with other chemicals. It is hoped that toxicologists who intend to service ERA, take the chemical aging considerations of this chapter to heart in refining their work.

References

Alexander, M. (1995) How toxic are chemicals in soil? *Environmental Science and Toxicology* 29:2713–2717.

Alexander, M. (2000) Aging, bioavailability, and overestimation of risk from environmental pollutants. *Environmental Science and Toxicology* 34:4259–4265.

Allard, P., Fairbrother, A., Hope, B.K., et al. (2009) Recommendations for the development and application of wildlife toxicity reference values. *Integrated Environmental Assessment and Management* 6:28–37.

Appell, D. (2001) The new uncertainty principle. http://sciam.com.2001/0101issue/0101scicit1.html 16 Dec 2000.

Bustnes, J.O., Yoccoz, N.G., Bangjord, G., Polder, A., & Skaare, J.U. (2007) Temporal trends (1986–2004) of organochlorines and brominated flame retardants in tawny owl eggs from northern Europe. *Environmental Science and Technology* 41:8491–8497.

Chard, B.K., Doucette, W.J., Chard, J.K., Bugbee, B., & Gorder, K. (2006) Trichloroethylene uptake by apple and peach trees and transfer to fruit. *Environmental Science and Technology* 40: 4788–4793.

Corsolini, S., Sara, G., Borghesi, N., & Focardi, S. (2007) HCB, p,p'-DDE and PCB ontogenetic transfer and magnification in bluefin tuna (*Thunnus thynnus*) from the Mediterranean Sea. *Environmental Science and Technology* 41:4227–4233.

Dietz, R., Bossi, R., Riget, F.F., Sonne, C., & Born, E.W. (2008) Increasing perfluoroalkyl contaminants in east Greenland polar bears (*Ursus maritimus*): A new toxic threat to the arctic bears. *Environmental Science and Technology* 42: 2701–2707.

Gallo, M.A. (1996) History and scope of toxicology. In: Klaassen CD, Amdur MO, Doull, JD (eds.) *Casarett and Doull's Toxicology, The Basic Science of Poisons*, 5th edn. McGraw-Hill, New York, USA.

Kendall, R.J. & Smith, P.N. (2003) Wildlife toxicology revisited. *Environmental Science and Technology* 178–183.

Kolluru, R. (1996) Health risk assessment: principles and practices. In: Kolluru, R., Bartell, S., Pitblado, R., & Stricoff, R.S. (eds.) *Risk Assessment and Management Handbook*. McGraw-Hill, New York, NY, pp. 10.3–10.59.

Lewis, R.W., Billington, R., Debryune, E., Gamer, A., Lang, B., & Carpanini, F. (2002) Recognition of adverse and nonadverse effects in toxicology studies. *Toxicologic Pathology* 30:66–74.

Marchant, G.E. (2003) From general policy to legal rule: Aspirations and limitations of the precautionary principle. *Environmental Health Perspectives* 111:1799–1803.

McNabb, F.M.A., Larsen, C.T., & Pooler, P.S. (2004) Ammonium perchlorate effects on thyroid function and growth in bobwhite quail chicks. *Environmental Toxicology and Chemistry* 23:997–1003.

Morrissey, C.A., Albert, C.A., Dods, P.L., Cullen, W.R., Lai, V. W-M., & Elliott, J.E. (2007) Arsenic accumulation in bark beetles and forest birds occupying mountain pine beetle infested stands treated with monosodium methanearsonate. *Environmental Science and Technology* 41:1494–1500.

Sample, B.E., Opresko, D.M., & Suter, G.W. (1996) Toxicological benchmarks for wildlife: 1996 Revision. ES/ER/TM-86/R3. Oak Ridge National Laboratory, Oak Ridge, TN, USA.

Schroeder, H.A., Mitchener, M. (1975) Toxic effects of trace elements on the reproduction of mice and rats. *Archives of Environmental Health* 23:102–106.

Sonoda, H., Kohnoe, S., Yamazato, T., et al. (2010) Colorectal cancer screening with odour material by canine scent detection. *Gut* doi:10.1136/gut.2010.218305.

Tang, J., Robertson, B.K., & Alexander, M. (1999) Chemical extraction methods to estimate bioavailabiloity of DDT, DDE, and DDD in soil. *Environmental Science and Technology* 33:4346–4351.

Tannenbaum, L.V. (2001) What's so bad about weight loss, blood chemistry effects, kidney toxicity, etc. in a modeled ecological receptor? *Human and Ecological Risk Assessment* 7:1765–1767.

Taurog, A. (1996) Hormone synthesis: Thyroid iodine metabolism. In Braverman, L.E., Utiger, R.D. (eds.), *Werner and Ingbar's The Thyroid*, 7th ed. Lippincott-Raven, Philadelphia, PA, USA, pp. 47–81.

Tickner, J., Raffensperger, C., & Myers, N. (1998) *The Precautionary Principle in Action: A Handbook*, 1st edn. Science and Environmental Health Network.

Troisi, G.M., Bexton, S., & Robinson, I. (2006) Polyaromatic hydrocarbon and PAH metabolite burdens in piled common guillemots (*Uria aalge*) stranded on the east coast of England (2001–2002). *Environmental Toxicology and Chemistry* 40:7938–7943.

Umbreit, T.H., Hesse, E.J., & Gallo, M.A. (1986) Bioavailability of dioxin in soil from a 2,4,5-T manufacturing site. *Science*. 232:497–499.

US ACHPPM (2000) Standard Practice for Wildlife Toxicity Reference Values, Technical Guide No. 254. US Army Center for Health Promotion and Preventive Medicine.

US ACHPPM (2007) Wildlife Toxicity Assessment for Perchlorate. Final Report, US ACHPPM Document No. 87-MA02T6-05D. US Army Center for Health Promotion and Preventive Medicine.

US EPA (1989) Risk Assessment Guidance for Superfund. Volume I: Human Health Evaluation Manual (Part A), Interim Final. EPA/540/1–89/002. Washington DC: US Environmental Protection Agency.

US EPA (1995) The Use of the Benchmark Dose Approach in Health Risk Assessment. Office of Research and Development. Washington DC: EPA/630/R-94/007. Washington DC: US Environmental Protection Agency.

US FWS (1964) Pesticide-wildlife studies, 1963: a review of Fish and Wildlife Service investigations during the calendar year. FWS Circular 199. US Fish and Wildlife Service.

US FWS (1969) Bureau of sport fisheries and wildlife. Publication 74, pp. 56–57. US Fish and Wildlife Service.

Von der Recke, R. & Vetter, W. (2007) Synthesis and characterization of 2,3-dibromopropyl-2,4,6-tribromophenyl ether (DPTE) and structurally related compounds evidenced in seal blubber and brain. Environmental Science and Technology. 41:1590–1595.

Voorspoels, S., Covaci, A., Lepom, P., Escutenaire, S., & Schepens, P. (2006) Remarkable findings concerning PBDEs in the terrestrial top-predator red fox (*Vulpes vulpes*). *Environmental Science and Technology* 40:2937–2943.

Wolff, J. (1998) Perchlorate and the thyroid gland. *Pharmacological Reviews* 50:89–105.

5 Risk characterization versus site ecological assessment: Old and new

The Superfund risk assessment processes we know of, one for human health (US EPA 1989a) and one for ecological receptors (US EPA 1989b, 1997, 1998) are remarkably similar in design. They both have their roots in the paradigm set forth by the National Academy of Sciences' (National Resource Council) landmark contribution to the health assessment field, known colloquially as "the Red Book" (NRC 1983). For our continued discussion of ERA topics, it matters not that the Red Book is squarely oriented to human health concerns; for all living creatures, an effective strategy for evaluating the potential for health impacts to arise involves the same essential steps. The naming of the steps may vary from agency to agency, from year to year, or by receptor type (human, animal, or plant), but empirically the risk assessment process that will undoubtedly always remain in place to lead us on to the paradigm's risk management element, are hazard identification, exposure assessment, dose–response assessment, and risk characterization. Based on the prior chapters, the attentive reader should observe a great difficulty – we're not fitted to characterize ecological risk. How could we be? We recall that there is no expression for ecological risk and that no true ecological risk assessment for a contaminated site has ever been conducted. The meager assessments that we *are* capable of (and that we might want to believe are risk assessments) evaluate chemicals singly and almost never put us directly in touch with the receptors that we aim to assess. For those still unconvinced and who would prefer to debate the point, their challenge would be to locate among their office collection of ERAs just one that articulated the probability (e.g., 5%, 11.2%, 32%) that a receptor would develop a toxicological effect. "Risk Characterization" sections of ERAs will readily and primarily report on HQs, but HQs are not expressions of risk (US EPA 1989a; Kolluru 1996). With reported HQs, we are not achieving what risk characterization sections of ERAs are after. If there is blame to lay for this great difficulty, it is in the terminology of US EPA ERA guidance documents. It is glaringly obvious today as Step 7 of the in-place 8-step process (US EPA 1997), but to get to the roots

Alternative Ecological Risk Assessment: An Innovative Approach to Understanding Ecological Assessments for Contaminated Sites, First Edition. Lawrence V. Tannenbaum.

of this matter, we should rightfully return to the EPA's founding ERA guidance document (US EPA 1989b), superseded over a decade ago by more developed guidance from that agency. Excerpted from Section 6.6 of that document in a section entitled "Characterize Risk or Threat", we have the following instruction:

> In characterizing risks or threats to environmental receptors associated with Superfund sites, the analyst should try to answer the following questions:
>
> What is the probability that an adverse effect will occur?
> What is the magnitude of each effect?
> What is the temporal character of each effect (transient, reversible, or permanent)?
> What receptor populations or habitats will be affected?

It can be argued that all of the above questions leave us wanting, but importantly for our purposes here, it is the first question that asks for information we have never been able to supply. That this guidance document and those that have come online since 1989 do not instruct on how to assimilate the information to allow for expressing the probabilities of adverse effects occurring, is one thing. To not be able to arrive at the end of the ERA process because we are unable to characterize risk, is another.

It may appear to be back-peddling, but to come to grips with our inability to characterize ecological risk and to discover a path forward, we must ask a rudimentary question: Why does one conduct a risk assessment? Variable answers to this can be expected, reflecting how the question is received. A short, technically correct, and not necessarily flippant answer would be "in order to find out how much risk there is at a site." A legitimate (and still not necessarily flippant) answer from a contractor might be "So that I can get paid, and hopefully draw in more work for my company." Let us work with the first answer though, to draw out a critical distinction. An expanded form of that answer would be: "in order to establish if risk is unacceptable, so that in the event it should be found to be so, efforts can be taken to forestall health effects from coming due." The reader should appreciate that this answer relates to human health risk assessment only. To dramatize the point, we will review a scenario that has been replayed countless times in Superfund history.

A routine groundwater monitoring/sampling round for a township identifies a few detections of the organic solvent, trichloroethene (TCE) in two monitoring wells. The township's health department is consulted and hurriedly calls for a re-sampling. TCEs again appear in the new samples, triggering a broader investigation. In just a few weeks it has become clear that there is a TCE plume advancing slowly through a sector of the township, and that for some period of time, residents have been drinking and showering with TCE-tainted water. The township residents are all placed on an alternate water supply as the investigation takes hold. A baseline risk assessment (BRA) is done for the affected homes, businesses, and schools, and the BRA considers, of course, what risks would pertain to an array of receptors (resident, teacher, worker) were they to continue using the contaminated water. The BRA finds unacceptable cancer risk and non-cancer hazard for several receptors and for several routes of exposure. The story, in part, may end here because we are not familiar with the practice of following up, months or years later, with individuals who had once been chemically exposed at contaminated sites. This is unfortunate because seemingly we should want to know if our interventions (here, taking the tainted water off-line) warded off health effects from coming due in these very individuals, a critically important point.[1] The express purpose of interrupting exposures, as in providing an alternate water supply to the township people in this example, was not only done to ensure that from the time the problem was discovered, there would only be an untainted water supply for drinking and showering.[2] Most importantly, it matters not how long the TCE plume might have been contaminating the township's water supply. There would be purpose in conducting a BRA in any year because reasonably, new township residents can be expected to arrive at any point in time to begin playing out their individualistic versions of the well-considered 30-year exposure at a fixed address (US EPA 1989a). Whether openly stated or not, the objective of HHRAs at contaminated sites is to determine if risk is unacceptable, this in an effort to prevent health effects from coming due in the population that has already been exposed.

Within any Superfund-type program, it is the hope that situations are responded to soon enough to pre-empt health effects in humans from taking hold. As similarly aligned as the HHRA and ERA processes are (and have always been), it has never been an ERA's purpose to endeavor to forestall health effects from coming due within the lives of the ecological species we observe occupying a site on a chosen day or season. We must believe this. Our proof requires only that we consider certain temporal elements: the life spans of the ecological receptors that concern us, the time elapsing between a site becoming contaminated and being identified as requiring assessment (what we may term "site discovery"), and the time lapse from site discovery until an ERA is completed. The EPA's Superfund program was created because in the 1970s we were discovering multiple-decade old contaminated sites. The ecological receptors that lived at those sites from the

[1] Admittedly, the prospect of conducting cluster-type studies can be daunting. Nevertheless, such efforts could demonstrate that despite successful remediation efforts, chemical exposures to site residents prior to cleanups taking place had been too high, unfortunately triggering disease years later.

[2] For the record, where we do not conduct health screening follow-ups, and where we more summarily assume that everyone is fine, our take on site close-outs is that "we" (usually the government) acted on the contaminated site matter expeditiously so as to forestall health effects from arising.

time they first became contaminated are no longer with us; they expired decades ago as they should have, obeying the natural biological order of the universe. Conceivably the receptors might have died from their chemical exposures at these unattended sites, but conceivably too, they lived out their full life spans unimpeded. We are certainly not going to investigate why, years ago, a field mouse, songbird, or raccoon at a contaminated site developed a chronic condition. If we consider just the time lapse from site discovery until an ERA is completed, we will observe some 7 to 10 years passing. Even this chunk of time is greater than the life spans of the receptors that commonly draw our attention.[3] What risks, then, are we trying to characterize at a site that has been contaminated for 35 years and has just now (say, at the time you are reading this chapter) had its ERA sent out for review? Arguably it is too late to be characterizing risk at a site that has taken its shots, decades of them in fact, at the resident biota. We cannot possibly intervene fast enough to rescue an ecological receptor from succumbing to the exposures it has already experienced. The upshot of this discussion is that our inability to characterize ecological risk doesn't really pose a complication for us; we won't find ourselves in a situation where we conclude that ecological risk is unacceptable and where we are prompted to devise a plan for rectifying or ameliorating the contaminated site condition.

The worrywarts would paint the picture an entirely different way. Conceivably, every site functions as a contaminant sink. According to this thinking, all ecological species that encroach on contaminated sites die, and must be investigated. Since sites have been contaminated for decades, they have been killing off all ecological species in uninterrupted fashion for all of that time. According to this thinking, we should not expect to see any life remaining. Sites though, to our perception, host an appreciable array of species and species representatives, or more correctly, a sufficiency of these; after all, how many elk should we expect to see at a 5-acre site? The worrywarts have an explanation for the presence of biota at the site as well; the numerous readily observable animals are actually migrants from non-contaminated areas that lie beyond the site boundary. Tenaciously holding onto the sink argument, the worrywarts conclude that over the decades, some mechanism has evolved for contaminated sites projecting the look and feel of pristine ones. If we suppose that the worrywarts are correct, the level-headed ERA practitioner should come to the conclusion that there is nothing to worry about despite the lethal character of sites. Why should we be bothered if nature has evolved a means for populating sites with biota although this evolutionary process has been and remains to be triggered by a chemical presence that limits the forms that can live at sites? The hypothesized sophisticated workings of sites that allow them to boast a vibrant ecology may be atypical and poorly understood, but these

[3] The respective approximate average life spans of red fox, squirrel, raccoon, Eastern cottontail, mourning dove, robin, barn swallow, tufted titmouse are: 1.5, 6, 3, 1.25, 7–10, 1.4, <4, 2.1 years.

are considerations that lie beyond the scope of ERA guidance and ERA tasks. It would be more appropriate to ask why anyone is bothered by the fact that a site regularly eradicates its biota, when it conveniently also has either stepped-up the biological processes of reproduction or immigration in successful efforts to replenish the losses. Importantly, we should not fall prey to the hypothesized ecological site operations, and in response to these, strategize to collect the information that could bear on the identification of compensatory processes that can account for the persistence of biota. The best fix for the hopeless theorizing would be to come to grips with reality. Superfund-style ERA efforts never dabble with sophisticated science, and certainly not when the HQ is held in such high esteem. Sophisticated field-based analytical efforts can't take hold for reasonably legitimate reasons anyway, such as limited project budgets and aggressive timetables for advancing sites to meet remedial program milestones that lead up to site close-outs.

The above discussion leads to a complication – a "risk characterization Catch-22" of sorts. Many ERA practitioners are ardent supporters of the sink argument. If they believe that every site is fooling us with its evident display of site-reared and site-supported biota, then dabbling with HQ computation should be perceived as a waste of time. But at contaminated sites with computed HQs < 1, we (i.e., the worryworts) are seemingly still suspect that everything is being killed off. With all of the uncertainty associated with HQs (discussed in the previous chapter), we cannot allow HQ outcomes to determine when the sink argument is to be switched on (HQ > 1) or off (HQ < 1).

At this juncture we would do well to review risk characterization in HHRA in support of a comparative analysis with ERA. We will do this in a novel fashion, presenting a compilation of data excerpted from the US EPA's CERCLIS database that the EPA has itself never reviewed or contemplated. We begin with the recognition that the overwhelming majority of HHRAs involve at least one or a few chemicals that are carcinogenic and one or a few that are noncarcinogens.[4] Consequently, most HHRA risk characterization sections report a site's propensity for eliciting two effects types: cancer outcomes (as the incremental lifetime cancer risk; ILCR) and non-cancer outcomes (as chemical-specific HQs or a cumulative hazard index, HI). Further, there are four pairings of outcome possibilities, and these are depicted in Fig. 5.1, constructed after a review of Superfund Records of Decision (ROD) of an 8-year period, with the RODs reporting on nearly 1000 contaminant exposure scenarios.[5] In many instances there were several scenarios evaluated per ROD, with the scenarios involving multiple environmental

[4] With a fair number of chemicals possessing the capability to both induce cancer and cause systemic effects, it should come as no surprise that so many HHRAs are dual-sided.

[5] Examples include incidental soil ingestion while gardening, inhalation of volatile organic compounds while showering, dermal contact with soil while performing housing construction duties.

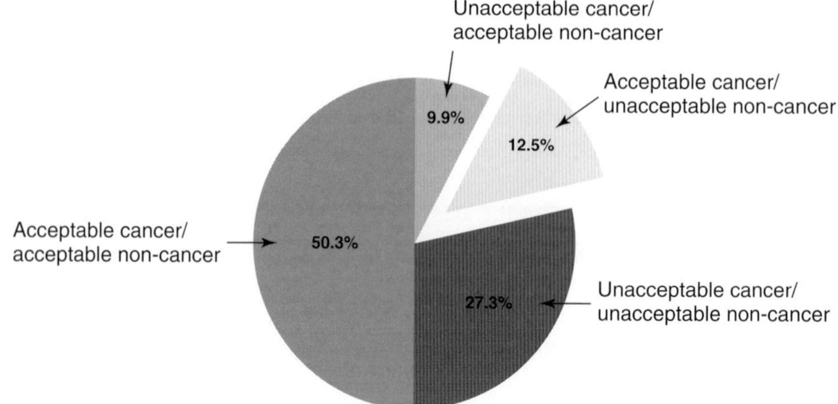

Fig. 5.1 Categorization of human health risk assessment outcomes. The human health risk assessment outcome of "acceptable cancer (risk) and unacceptable non-cancer (hazard)" is particularly rare, so much so that risk managers are not well prepared to deal with it. (Figure 5.2 informs that human health HI exceedances of 1 are rather inconsequential.) In ERA, HQs (frequently exceeding 1) are always apparent because there is no cancer assessment outcome to shield them. (adapted from Tannenbaum et al. 2003. Reproduced with permission of Taylor & Francis.)

media. Parenthetically we initially stumble across a piece of very good news, namely that 50% of the time, sites pose no human health concerns regardless of endpoint. Let us then deal with the remaining half of our pie chart. We find that approximately one quarter of all cases present "unacceptability" for both cancer and non-cancer. Since there is no debate that cancer risks greater than 1E-04 (an anticipated site-posed cancer incidence of one in ten thousand) trigger remedial action, we can readily look past this pie sector, i.e., there is no great need to closely examine what chemical(s) contributed to the unacceptable HQ/HI condition in the evaluated scenarios. Unacceptable hazard presents itself in one other pie chart sector. For about one-eighth of all cases, cancer risk is acceptable whereas the non-cancer hazard paired with it is not. This highlighted portion of the pie chart could concern us greatly.

Hazard, whether for HHRA or ERA, is not a probability measure and there are no clear rules for reacting to this pie chart sector "cancer/non-cancer arrangement." The complication for risk managers in deciding how to responsibly apply a remedial action when HIs are >1 is exacerbated with the information presented in Fig. 5.2, a histogram of HI magnitudes corresponding to the nearly 1000 scenarios considered. To the right of the first two histogram bars we observe a literal handful of HIs of a magnitude we might refer to as sizeable, with some of the histogram columns (verified through follow-up research) reflecting singular cases.

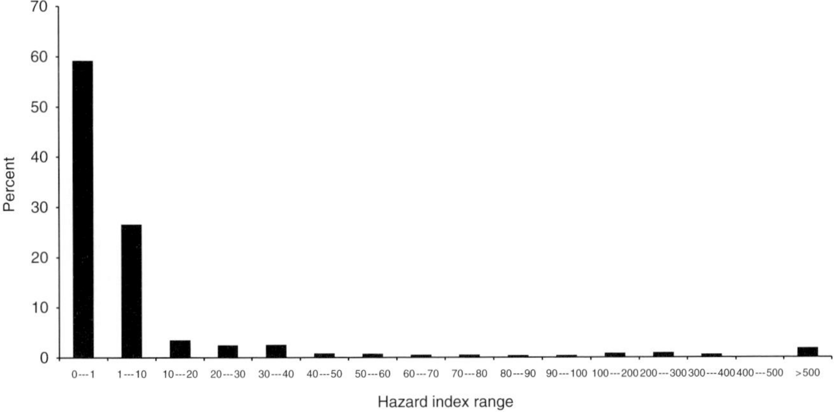

Fig. 5.2 Histogram of the magnitude of human health HIs. In some 85% of cases or more, HI are less than 10, with most less than 1. (Data from Tannenbaum et al. 2003.)

Additionally, nearly all of the researched singular-case high HIs, reflect instances of an organic species (e.g., PCBs) that presented itself to the receptor(s) as *pure product*, an example of this being a puddle of TCE observed on the ground. In such an instance, sampling and analysis to render an exposure point concentration for use in risk assessment work is completely out of place, for a puddle of TCE is not a *bona fide* environmental medium (as are soil, sediment, surface water, air, or groundwater) to be assessed in Superfund-type work. For all intents and purposes, consideration of nearly all of the HIs to the right of the two first two histogram bars falls away, fortunately helping us avert the pitfall of lending credibility to the excessive magnitude values. HQ and HIs do not conform to a linear scale, but overlooking this, one could think that humans at some of the sites are consuming, inhaling, or dermally contacting chemicals at 100 or more times the safe dose. More than anything, the excessive magnitude HIs demonstrate the frailties of the HQ construct. Screamingly high chemical concentrations give rise to highly non-representative (and worrisome) projections of chemical exposure and disease potential. Several critical points distill from the review of the pie chart and histogram:

- Human health HIs rarely "fail" and when they do, it is only by a drop (i.e., values less than 10).
- Should the HI exceed 1 at a site, this will almost always be paired with cancer risk that is also unacceptable. When this occurs, it matters not what the HI magnitude is, for all attention – down to the nature of the remedial strategy to be invoked – will be thrown towards the unacceptable cancer finding.

- Combining the two previous points, it is a rare event that only non-cancer is found to be unacceptable. Additionally, when this rare event arises, we are almost always dealing with HIs of a most minimal magnitude (perhaps just 3 or 4). This situation can be most unsettling for remedial managers who likely have never encountered it.[6]

In summarizing our understanding of risk characterization in HHRA, we might apply some admittedly twisted and figurative speech for shock value in cementing a critical point: "Thank G-d for cancer." "Risk" characterization, as a health assessment component of a larger process, is a legitimate term to use because cancer assessment outcomes are indeed probabilities. Invariably, at least one chemical carried through each HHRA is classified as a carcinogen, and the reporting of one or more ILCRs upholds the usage of "risk" characterization as an assessment report section heading. If HHRA only had the capability to report non-cancer outcomes, it would be in the same boat as ERA, and the section heading corresponding to the HHRA process's final step should appropriately be "Hazard Characterization" as opposed to risk characterization. In fact, should a given Superfund-type site only "screen in" systemic (i.e., non-cancer causing) toxicants as contaminants of potential concern, "Hazard Characterization" would be the most appropriate and serviceable section heading to use. With cancer not being an ecologically relevant endpoint, ERAs can at best report HQs, that we will recall are unitless values and certainly not probability expressions. For our purposes, it is important to see that it is wishful thinking to report ecological HQs in a "risk" characterization treatment. Doing so brings along the unfortunate effects of duping ERA practitioners into believing they are indeed characterizing risks, and making it ever-more unlikely that ecological assessments will ever be improved.

A related point that follows from the above analysis is that HQ limitations, commonly applicable to HHRA and ERA, are far less noticeable in the former. In HHRA, HQs are able to hide behind ILCRs since the latter speak to the prominent driving force of our (human) health concerns, cancer. Only a step short of all-out paranoia, we are bent on knowing the likelihood of people developing the scourge that is this horrid disease. Curiously, we may even go so far as to list HQ method limitations in HHRA uncertainty sections (and indeed this is what should be done), but this needn't mobilize us to develop a bettered assessment scheme for systemic effects. In ERA though, HQs are fully visible because they have no other computation or reporting format to mask them. Further, where we list HQ method limitations in the uncertainty sections of ERAs, the ERA process

[6] For a site remedial strategy to be negotiated with a state agency, the author had the experience of having a senior Superfund manager ask him to "see about making some more risk," referring to seeing if cancer risks at a site could be recomputed to exceed the 1E-04 trigger, or if HIs could be made higher.

bears the brunt of these as it should. All things considered, we arrive at the understanding that ERA's standard reporting format, the HQ, is insufficient for our needs. The added complication of ERA HQ magnitudes tending to be pronounced, only serves to aggravate matters. Excessive magnitude HQs, perhaps as low as 20 (NOAEL-based) have been described in the literature as being unrealistically high and toxicologically impossible (Tannenbaum et al. 2003).

The foregoing discussion really begs the question of why the so-named "Risk Characterization" section of ERAs, as a reporting format fixture, hasn't been challenged or replaced by a more appropriate term. Before putting attention to this, we should review the sort of information that we can expect to find in risk characterization sections – aside, of course, from the HQs that have been amply discussed in this and the earlier chapters.

- Frequently intertwined with HQ presentations are statements concerning the spatial relevance (or more correctly here, the spatial irrelevance) of the considered receptors. Risk characterization sections will acknowledge that the home ranges and foraging ranges of several, if not most receptors, exceed the site size by a great deal, perhaps by tens or hundreds of times. Pursuant to such acknowledgment, the text will typically then indicate that for such receptors, there are no anticipated health effects. The astute document reviewer will note and hopefully comment to the effect that if it was known prior to HQ computation, that the wide-ranging receptors were not spatially relevant for the site in question, the receptors should not have been selected in the first place. For our purposes, we are challenged to see how such reporting qualifies as risk characterization information. Perhaps in an unintended way, the reporting does so qualify – by acknowledging the relatively miniscule site size relative to receptor movements, the text is effectively communicating that there is no risk to characterize because there were no valid receptors.
- Commonly, risk characterization sections will acknowledge that a site is of low-quality habitat and is not expected to draw the receptors that were considered in the ERA. This acknowledgment, that may have been alluded to earlier in the ERA and possibly as early as the Executive Summary, gets us to the difficulty we encountered in the previous paragraph – if before any ERA techniques were applied (i.e., HQ computation or otherwise) it was known that the site is mostly asphalted, that it abuts a section of highway, or is littered with construction and demolition debris, we should again be left wondering why ERA efforts of any order proceeded. It is not an ERA's place to assess the imaginary case of a site offering suitable, if not splendid habitat, if in actuality it does not.
- When discussing vegetation effects, many a risk characterization section will report a series of toxicological benchmarks for terrestrial plants (i.e., laboratory-based soil concentrations that are known to contravene plant growth and

plant health) that were exceeded. Such reporting is then followed by statements to the effect that the site (perhaps other than an occasional patch of non-grassed soil) is well-vegetated and bears no signs of stressed vegetation. This reporting style begs the question of why toxicological benchmark screening proceeded altogether. Presumably well before the screening effort occurred, investigators had an awareness from their site walk-throughs and field sampling efforts, that overt signs of vegetative impact (barren patches of soil, stunting, discoloration) were absent. In that common case where a site's plant community doesn't appear to be faltering in the least, there is no need for a plant assessment of any sort, let alone a risk characterization.[7] The reader should appreciate that an often-accessed compilation of plant toxicological benchmarks (Efroymson et al. 1997) prominently caveats its own utility.[8] Briefly, if the site is vegetated, there's no need to screen soil contaminants for their potential to elicit plant health effects. We might wish that all collections of eco-toxicological benchmarks for various forms (e.g., earthworms, benthic macroinvertebrates) similarly instructed users that contaminant screening is unnecessary when the various forms are present.

- Where toxicity testing was conducted as part of an ERA, we can expect risk characterization sections to supply an interpretation of the test outcomes. Here the reader is advised to recall Chapter 4's treatment of this subject. It is imperative that one understand that toxicity tests will not have been conducted with site organisms, but rather with commercially reared specimens that had no prior exposure to contaminants. Although the inclusion of a discussion on the toxicity testing conducted provides an information-added component to an ERA's risk characterization, the discussion's limited scope must be realized. Without exception, toxicity testing only comes to address the imaginary case of unsuspecting ecological species having been released anew to a contaminated environment. We don't find that mist-netted songbirds are ever let go at thicketed sites by good-natured environmentalists, and perhaps the only instance we know of man-transported biota is that of stocking lakes with fish. With regard to toxicity testing, we don't know of earthworms being brought to terrestrial sites for release, nor has it ever happened that benthic macroinvertebrates were deliberately hand-carried to contaminated site waterbodies and set out there to colonize the habitat.

[7] The reader is reminded that of all ecological receptors, terrestrial plants are the easiest to directly assess. Plants require no trapping and no special skills are required to assign a qualified overall statement of health.

[8] The cited reference cautions that where vigorous and diverse plant communities are supported, although chemical concentrations in field soils exceed one or more benchmarks, it is generally safe to assume that benchmarks are poor measures of risk to a site plant community.

We have established many pages before that there is no way to characterize ecological risk. Because this is so, each ERA being assembled will find itself on a collision course from its outset, as follows. A textual treatment articulating the degree to which risk is manifested at a contaminated site will need to be supplied in an ERA's final chapter. The information that will be presented there will follow the pattern of the above bulleted paragraphs. Largely the presented arguments fall into two categories – unfounded claims that risk is unacceptable, and statements such as have been typified here, that come to write-off (any) potential problems. In truth, the latter category should be seen as reflecting an embarrassingly unsophisticated and ineffective assessment process. Since it wasn't possible to measure or assess risk, as filler, risk characterization text will only describe (again) that the (investigative) work was done. Since it wasn't possible to measure or assess risk, the text will frequently explain that there wasn't a problem to investigate after all. Why there isn't more public outcry from ERA readers over having been duped into initially thinking that a given site had a chemical exposure problem, only to find that ERAs back off of such a position, is an excellent question. Seemingly our reticence stems from being excessively format-ridden, preferring to see that all ERAs conform to a strict reporting template and giving far less attention to the value of the material provided in a required ERA chapter or section. A related and equally valid reason is that non-regulatory parties are afraid to buck the process. Before an alternative approach to the conventional reporting at the process terminus can be offered, we must allow ourselves to see that adhering to the conventional assessment scheme has gained us no ground. An amply vegetated site doesn't need to have its soil chemical concentrations reviewed to see if any might pose threats to plant life. A physically disturbed site that hardly supplies essential habitat or that is unlikely to lure receptors to it for this or some other reason, doesn't need a risk assessment or risk characterization. Having identified sediment chemical concentrations that exceed, even by many times, tabular values for the protection of sediment-dwelling and other aquatic species, is not "the ticket home" to conclude that aquatic receptors are at risk. It is certainly not "the ticket home" given (a) what the earlier review of such screening values brought to light, and (b) when exceeded screening benchmarks aren't shown to correlate with depauperate or skewed aquatic receptor populations.

Transparency in reporting would seem to be the alternative to replaying the risk characterization verbiage with which we have become so accustomed. At a minimum, an ERA section to replace "Risk Characterization", perhaps to be titled "Summary of Findings", would acknowledge that risk could not be computed or expressed. To gain traction with this modified reporting mechanism, the new section should first explain that given the limitations of the science, "Risk Characterization" is a misnomer. The Summary of Findings section should then overtly state that it is fully intended to supersede Risk Characterization, briefly explaining with boilerplate text that ecological risk is something that cannot be characterized. The section would then become more site-specific, indicating that because the likelihood of health effects or impacts arising in site biota cannot be expressed, other tasks were completed as a next best means of gaining a potential-for-health-effects-arising perspective. Barring the most unlikely of situations, where site visits reveal all-out death, doom, and destruction, all ERAs would predictably have their Summary of Findings sections commonly conclude, again with boilerplate language. They would state that based upon the information that was collected, along with constraints on more aggressively sampling and studying the site, and overall science limitations, there is no evidence to say that ecological receptors are at risk.

The suggested reporting format above is far from the best repair that ERAs could see. Importantly though, the suggestion permits truthfulness to be conveyed while still allowing for conventional and unfortunately ineffective lab, field, and desktop efforts to be presented. That is, one could still report (and read) about deer, hawk, and raccoon at a 3-acre site having been evaluated with a food-chain model that employed area use factors of 0.002, 0.001, and 0.008, respectively.[9] So long as the upshot of the HQ reporting wouldn't be that these receptors bear a potential for risk, there needn't be discussion at this late point about the meaning of HQs or that the species were poor modeling selections to begin with.

An alternative and much-enhanced concluding ERA section would make secure the pronounced point that in truth, a risk assessment wasn't needed. In actuality, it would be nearly impossible to maneuver such text into an ERA's concluding section, for this would effectively be communicating that all study was conducted under a false pretense. Thus, implementing this best alternative would necessitate a complete revamping of the ERA process. As will be discussed in Chapter 9, the goal is not to do away with assessments entirely, but to conduct assessments that are appropriate. Indeed there are such assessment schemes that do rise to the occasion.

The greatest impediment to adopting a notable shift away from present-day risk characterization treatments is undoubtedly our unwillingness to hear that risk cannot be characterized. We would do best then, to consider why the ERA powers-that-be so tenaciously hold onto the notion that risk *can* be expressed. It could be that a fundamental misunderstanding of what Superfund and other programs intend to elucidate, forms the crux of the matter. Superfund and other risk-based programs seek to know how much risk there is for a given environmental setting, and these programs do not merely want to know *that there is risk.* As it would appear, ERA practitioners are focused only on the latter information, and this poses its own difficulty. Establishing a receptor's level of risk enables appropriate cleanup values to be developed; simply knowing that there is risk (generally understood to

[9] Curiously, as he was writing about this point, the author just happened to be reviewing an ERA with these particulars.

connote "unacceptable" risk) does not enable such cleanup value derivation. The magnitude of the difficulty must be appreciated. At an initial level, being copacetic with determining (in some manner) that there is risk, primes us to either move for remediating chemical concentrations down to proverbial site background conditions, or for back-calculating HQs to determine chemical-specific concentrations that equate to a HQ of 1. We should recognize that neither of these approaches are very much grounded. Chemical concentrations, if they are troublesome at all, probably don't need to be completely obliterated. HQs are not linearly scaled as we well know, and there is no legitimate basis for the back-calculating. Although back-calculation may be a mainstay in HHRA, there is an apples-and-oranges comparison here. In HHRA the calculations can be allowed because of the primacy we place on human health protection, something most understandable. Additionally, where back-calculation may proceed for non-cancer assessments in HHRA, in the overwhelming majority of cases, there'll be no practical need to have the computations anyway. As was illustrated earlier, attending to any instances of unacceptable *cancer* risk will all but relegate HQs > 1 to an afterthought.

An inability to express ecological risk, or more correctly vis-à-vis this chapter's topic, to characterize risk, is no small matter. We might consider how often we have been exposed to criticism of the ERA process, with specific criticism directed at the inability to characterize risk. We might find that other than this chapter, no other parties have elected to broach the subject. To the author's way of thinking, if this topical area hasn't garnered much attention, or any attention at all, we should try and explain why this is so. Bringing the sheer inability to characterize ecological risk out into the open might constitute our best chance of having the powers-that-be come to recognize how ineffective the ERA process is, and with that recognition comes the potential for seeing methodology reform. For now, it is plausible to suggest that ERA practitioners aren't sufficiently motivated to challenge the system, feeling that any expressed criticism of the in-place process will only fall on deaf ears. Alas, this suggestion might have us giving too much credit to the ERA practitioner. Perhaps ERA practitioners don't have the sense that ERA risk characterization sections are lacking altogether! Where this latter case should happen to accurately capture the sentiment, the source of the difficulty may simply lie in a misunderstanding of the capability of the HQ, namely the belief that HQ magnitudes articulate levels of (unacceptable) risk.

References

Efroymson, R.A., Will, M.E., Suter II, G.W., Wooten, A.C. (1997) Toxicological Benchmarks for Screening Contaminants of Potential Concern for Effects on Terrestrial Plants, 1997 Revision (prepared for US Department of Energy) Oak Ridge National Laboratory. ES/ER/TM-85/R3.

Kolluru, R. (1996) Health risk assessment: principles and practices. In: Kolluru, R., Bartell, S., Pitblado, R., Stricoff, R.S. (eds.) *Risk Assessment and Management Handbook*. McGraw-Hill, New York, NY, pp.10.3–10.59.

NRC (1983)(National Research Council) *Risk Assessment in the Federal Government: Managing the Process*. National Academy Press, Washington, DC, USA.

Tannenbaum, L.V., Johnson, M.S., & Bazar, M. (2003) Application of the hazard quotient in remedial decisions: A comparison of human and ecological risk assessments. *Human and Ecological Risk Assessment* 9:387–401.

US EPA (1989a) Risk Assessment Guidance for Superfund. Volume I: Human Health Evaluation Manual (Part A), Interim Final. Washington DC: US Environmental Protection Agency. EPA/540/1–89/002.

US EPA (1989b) Risk Assessment Guidance for Superfund. Volume II, Environmental Evaluation Manual, Interim Final. EPA/540–1–89/001. US Environmental Protection Agency.

US EPA (1997) Ecological Risk Assessment Guidance for Superfund: Process for Designing and Conducting Ecological Risk Assessments, Interim Final. EPA/540-R-97–006. US Environmental Protection Agency.

US EPA (1998) Guidelines for Ecological Risk Assessment. Washington DC: Risk Assessment Forum. EPA/630/R-85/002F. US Environmental Protection Agency.

6 Case study: Problem formulation versus making problems for yourself

As the saying goes, recognizing that one has a problem is the first step towards getting the problem addressed and even corrected. A person who imbibes alcohol on frequent occasions and who drinks larger quantities than we know to be the norm, has an addiction that might get out of control if it has not already. This much we know from the experts who counsel and tender medical care to those so afflicted. Coming to the recognition that one has a medical problem like alcohol abuse might take the form of a family member confronting the person in need about the behaviors he or she has observed. The problem being brought out into the open might occur at a physician's office, with the latter detecting classic tell-tale signs of disease upon patient examination. In coming to recognize that one has a medical issue in need of treatment and repair, perhaps it is best when the afflicted individual, without any prompting, acknowledges the problem on his own. Be it a medical issue, or a contaminated site issue wherein ecological receptors stand to be stressed or impacted, assigning a "problem" label constitutes a pivotal juncture. Establishing that there is a *bona fide* problem directs the necessary skill sets to address it. Without a defined problem, we cannot know what we are trying to correct, or have a sense of how extensive the correction needs to be.

For sites that are to be considered within a Superfund framework, the ERA process has formatted, in a most stereotypical way, the means by which a problem is identified. To set the ERA process in motion and to conclude there is a problem suited to be formally articulated and dealt with in a "problem formulation" statement and step (US EPA 1997), all that is needed is the knowledge that an environmental medium has a contaminant load. If PCBs have spilled onto soil, the formulated problem will be one of mammals incidentally ingesting PCBs when occupying the affected area, with reproduction to potentially be seriously offset. With nary an exception, the problem formulation in this specific instance will almost assuredly be honed to consider mink (*Mustela vison*) as that distinct species to be rendered reproductively compromised – for every ecological risk assessor today and his brother has been primed to robotically and unimaginatively

Alternative Ecological Risk Assessment: An Innovative Approach to Understanding Ecological Assessments for Contaminated Sites, First Edition. Lawrence V. Tannenbaum.

regurgitate that mink is the most reproductively sensitive mammal with regard to PCBs exposure. (Perhaps this is true. No matter if this is true. Please read on.) If one or more pesticides have been detected in the sediments of a river reach, the formulated problem will be one of behaviorally and reproductively impacted fish populating the area as a consequence of the pesticides influx. If substantial quantities of lead are freely available on the ground surface at a mine tailings site, the formulated problem will be one of resident birds standing to become neurologically impaired if not killed off outright, from the high dietary lead exposures we suppose are taking place.

Contaminants can unquestionably disrupt a site's ecology, and they can compromise any number of a receptor's essential biological functions. Perhaps knowing only that a chemical release has occurred should be reason enough to trigger the ERA process. By the same token, there are legitimate reasons not to let a site's chemical signature trigger the problem formulation step that so many of us are primed to see. One reason to be hesitant about launching into the 8-step process armed only with chemical signature information is that we know of so very few contaminated sites where manifested ecological effects *of import* were evident. This point necessitates additional attention. While we are able to cite several well-known examples of benthic macroinvertebrate community structure or abundance having been unmistakably modified due to contamination, such as that of Commencement Bay in Washington state (Becker et al. 2009), the ramifications of the effects may not be known, or there may be no ramifications. In the case of Commencement Bay, the EPA may have deemed the shifted benthic macroinvertebrate community to be an "effect of concern", but we should be wary of the information that factored into that position, assuming such existed. A legitimate basis for the developed position would be the knowledge that shifted macroinvertebrate communities translate into perturbed fish populations, and perhaps too, perturbed populations of piscivores, but do these translations occur? It's highly questionable that such information exists, and it's questionable as well that such information constituted EPA's thinking or proof regarding Commencement Bay. In the absence of a reason for a shifted benthic macroinvertebrate community being a bad thing ecologically (i.e., truly being "of import"), the utility of advanced studies to ultimately rectify the community structure are seriously questioned. Was Commencement Bay's overall productivity compromised by the benthic community shift, and if it was, what degree of reduction in productivity would absolutely necessitate action? Were there fewer fish in the bay as a result? While several fish species in Onodaga Lake in upper New York State had unhealthy mercury levels vis-à-vis an HHRA context, fish populations there weren't depleted. Lest we lose hold of the chapter's initial discussion, we should note that sites like Commencement Bay or Onodaga Lake are not really relevant to this book's scope. Most (aquatic) sites are not 4.5 square miles or 12 square miles in size, as are Onodaga Lake and Commencement Bay's nearshore/tideflats, respectively. Before we re-orient our discussion, we must decide if our attention to contaminated sites is driven by having contamination in place that we'd rather not have, or because we know that ecological damage has occurred. It is the latter, of course, that should drive our attention. Still, it might suit us well to briefly review the case of benthic macroinvertebrates in yet another consideration of scale. We should not be so brash as to think that organisms as small as benthic macroinvertebrates are unimportant, or to think that there is an official minimum size requirement before a living form can first be drawn into ERA consideration. Benthic macroinvertebrates are visible to the naked eye, but they are best described by the mesh size that retains them when sediments are filtered; they are retained by mesh sizes greater than 200–500 μm (about 1/64 of an inch), and they include insect larvae (the predominant form in freshwater aquatic systems), annelids, oligochaetes, crustaceans, and gastropods. If benthic macroinvertebrates constitute such a key focus in aquatic assessments, why don't we see the same emphasis placed on ants and other miniature soil-dwelling forms at terrestrial sites? Unless we have collectively weighed in with a risk assessment exclusion policy for insects at terrestrial sites (and I don't believe we have), we are seemingly overlooking a potentially critical component in terrestrial systems paralleling the role(s) served by benthic macroinvertebrates in aquatic systems. Perhaps contaminated soils at sites are responsible for only pollution-tolerant ant species populating them, or perhaps the soils are limiting to ants overall. Ants, like benthic macroinvertebrates constitute the base of food chains, and there should be just as much reason to give pause to the ecological consequences of shifts in ant population dynamics.

The truth might be that size does "matter", meaning that miniscule forms are unimportant, and only organisms that do not have to resort to being measured on a micron scale should garner our attention. Perhaps we make such a big deal of benthic macroinvertebrates because they tend to remain in their original habitat, and because we are fluent with distinctive feeding groups (such as shredders, filter collectors, grazers, and predators) and their pollution tolerances.

With a track record such as we have, we might stop what we are doing just long enough to contemplate how likely we think it is that a site we might begin assessing will pan out to have something sufficiently wrong ecologically, and to the point that intervention is needed. Another reason to defer is that each time we launch into the 8-step process, we know *a priori*, or we should know *a priori*, that the process, with nary an exception, will only lead us to the calculation of HQs and no farther. And HQs, excessive in magnitude as they might be, do not inform on sites actually bearing health-compromised forms. With these realizations (problems?) about the problem formulation step in mind, and a sense that we are following the 8-step process because we don't have the option not to, it would be appropriate to consider an alternative way of thinking. To get there, we must first realize that it can't always be that a chemical presence translates into a threat to site ecology. Our mediocre computations aside, we must also come to grips with our inability (other than what Chapter 9 has to offer) to ever know if ecological effects are, or have been, taking hold. Perhaps instead of defaulting each time to a chemical presence as the initiator of problem formulation (in any of the forms reviewed in Chapter 2), our energies should be channeled into scrutinizing the information we might be repeatedly and blindly accepting as laying a legitimate basis for ERA exploration. We really *do* have every reason to believe that the information available to us might not be enough to go on. If we find the notion of applying this enhanced scrutiny of information to be unappealing, our biases might be showing through. We should not be viewing every potential site as a property that needs to be made cleaner/more chemical-free. Given the track record of not finding health effects at sites at the population scale, we should seek out opportunities to recognize when our efforts might be pointless and wasteful and where our precious resources can be more sensibly applied. By way of example, bird losses might regularly occur at certain contaminated sites, and we might clearly know the specific offending chemicals involved. Cautiously applied reality checks though, as in considerations of scale (i.e., the meaningfulness of site size) might inform us that the losses might be only single-digit affairs. Seemingly our attention should not be turned to such site settings, which can undoubtedly be tolerated, and especially when hundreds of millions to greater than one billion North American birds are directly killed each year at the hands of human stressors. If anything could be threatening bird populations to calamitous levels, it would be the ongoing collisions of birds with man-made structures such as office buildings and wind turbines, in addition to other substantial losses through predation by feral and pet cats, and intentional and accidental poisoning (Loss et al. 2012). Why again are we bothering with crude desktop assessments for hypothesized species that are so sparsely distributed in nature that we can at best only anticipate a nest or two being present on a contaminated property?

In a very real sense, we have learned in the earlier chapters how to scrutinize site information, and thereby identify when chemical threats are sufficiently, if not entirely absent. Thus contamination might abound at a site, but attendant spatial realities critically influence the complexion of a site, and these are not to be ignored. We must decide what species concern us and if we have enough of them present to legitimize our attempts at assessment. It's not anyone's fault that contaminated sites tend to be small, that we haven't a whole lot of representatives of the species that concern us and that these species tend to cover lots of turf. And so let us return to those examples of would-be concerns with which we began this chapter.

- PCBs in soil and riverine sediments might have a definitive site footprint, but are mink associated with the site – recalling that mink have an average home range of about 100 acres (safely, and certainly not an underestimate; Burt &

Grossenheider 1980; Chapman & Feldhamer 1992; US EPA 1993; CH2M Hill 2001), and generally a linear home range at that in the case of riverine forms? If mink *are* a possibility, are there enough of them present that stand to be potentially endangered, to thereby facilitate embarking on an assessment?

- Pesticide-laden sediments might typify a river reach, but have we an adequate assessment scheme to bring us to a meaningful understanding of what effects the pesticides might pose to resident fish? To make for an indicting case, we should know how many fewer fish there need be in the affected reach relative to other non-affected reaches, but this information is probably unknown to us. The affected reach could have an altered benthic macroinvertebrate array, and this could negatively affect fish health. Unfortunately we must realize, there are no means for relating these two system elements (i.e., we don't know how much alteration of a benthic macroinvertebrate array needs to occur such that fish that feed on this level become nutritionally shortchanged). While efforts to employ models to answer such queries could proceed, we should have the sense that fish in our reach are nutritionally shortchanged to begin with. Finally, it's doubtful that fish behavior will be studied in the affected reach, and we must recall that there is no standard behavioral measure for fish *in situ*.
- Neurologically impaired birds are a distinct possibility at lead-contaminated sites, but blood lead concentrations, even extreme ones, don't demonstrate the impairment that we'd like to know of (as will become apparent later in this chapter). Without the necessary toxicological support, we only complicate matters by embarking on certain assessments. As an overarching consideration, how neurologically challenged could birds be at mine tailings sites if they are present in sufficient number to submit to mist-netting and subsequent blood draws and other examination (Sample et al 2011; Hansen et al. 2011)? In line with earlier considerations expressed on multiple occasions, due to site size, have we the anticipation that there are enough birds present to support an investigation at all. Most sites do not extend for tens and hundreds of acres, we are reminded.

This chapter's review of the problems associated with problem formulation removes itself from the most elementary case described at the start – that of a chemical presence in one or more site environmental media assumed to have the potential to wreak ecological havoc. In truth, the most substantive difference for the to-be highlighted case relative to the conventionally considered arrangement (of a contaminant being either embedded or dissolved within a medium to present itself as concentration) is that the contaminant takes the form of an independent particle (actually an unfathomable quantity of particles) sitting for the most part, atop the soil. We are, of course, speaking of spent lead shot pellets littering the ground surface in the fall zones of trap and skeet ranges. This review will hopefully accomplish considerably more than merely illustrating that a contaminant presence need not necessarily trigger ERA investigation. The intent is to raise consciousness, with the reader coming to appreciate that there isn't always a problem to pursue. Further, we stand to gain when we cautiously and correctly avoid those situations that do not call for ERA. Since various components of the ERA process are science-compromised, we should appreciate how much better off we are when we legitimately discover that we do not have to engage in them.

Going back some 50 years or more, the open literature considers the health effects of birds consuming lead particles left behind from recreational shotgun activities, principally hunting, and trap and skeet range use. Many species of upland and aquatic birds, through foraging activity, can come to incidentally ingest pellets of variable size. In the case of waterfowl, sifting through ponded sediments in pursuit of crustaceans and other dietary items can result in the consumption of relatively large shot pellets (perhaps size 4), the artifacts of hunting. Not surprisingly, the mass of lead within a single large pellet of this type, if allowed to dissolve mostly or entirely before being excreted, can produce lethality in a very short time. The phenomenon of lead pellet consumption by upland birds is notably different, and primarily because consumption events where they occur, are deliberate actions. Here we refer to that category of bird known as a grit ingester (i.e., a species that spends some portion of its time on the ground, and that with some frequency, ingests particles of rock or stone measuring from 1 to 4 μm in diameter). Grit ingesters retain the consumed particles in the muscular second portion of the stomach, the ventriculus (or gizzard). Presumably the behavior occurs to assist digestion; with one or more grit particles retained in the gizzard, the bird can contract the gizzard to rub the particles up against certain dietary items (seeds, chitinous insects, etc.) to facilitate a mechanical grinding action. It has also been suggested that true grit in the form of bits of sand or other small pieces of rock are ingested to supply essential nutrients such as silica. For potential ERA purposes, it matters not which reason is the correct one; it is sufficient to know that the behavior proceeds, and that grit ingestion behavior, in theory, can lead to bird death at trap and skeet ranges. Spent shot pellets are commonly size 7 1/2, 8, or 9, and are of a size similar to the preferred grit of many species. For decades it has been suggested that at the approximately 9000 non-military target shooting ranges in the US (US EPA 2001), upland birds could mistakenly ingest spent shot pellets as grit, to then present the same problem described earlier for waterfowl. For as long as a shot pellet, which may consist of 90% lead or even more, resides in the gizzard, the lead will dissolve to then be circulated in the bloodstream. As a neurotoxin, the lead will cause birds to become sluggish and as advanced toxicosis sets in, lethality will follow.

How will we decide if the availability of spent shot pellets to grit ingesters fuels problem formulation as per the ERA process? We might, in part, consider the magnitude of the lead pellet releases, for they will convincingly show that opportunities for birds to encounter pellets are unlimited. Let us work through an example to show this. There are no less than 400 pellets of the sizes just described in one charge fired from a shotgun. Twenty-five charges are fired in a round of trap or skeet, and reasonably, one who goes to a range for this type of recreation will probably shoot two rounds on a given day. We can assume that five individuals will recreate on a given day that a range is open for business.[1] We will assume that a range is open for 2 days/week, and for 9 months/year. Multiplying all of these terms together produces a figure in excess of 7 million pellets being deposited in a year's time. It goes without saying that with ranges having been in operation for multiple years, and with no reason for any of them to have been swept of their accumulations, billions of pellets litter the ground surface of a given fall zone. At this level of assembled information, we would be hard-pressed to deny that completed exposure pathways (pellet-to-bird) exist at all trap and skeet ranges. Exposure assessment however, must be given its due attention.

While pellets are not limiting, birds might be. It could be shortsighted then, to launch into a risk assessment of some order, or perhaps some 9000 of them, solely because of the undisputed availability of spent pellets. And so, how many birds might we assume are at risk from incidentally ingesting a spent shot pellet at a given trap and skeet range? – a fair question to pose. One approach to take when fielding this question is to consider fall zone size. Recreational shooters fire from fixed positions at ranges, some seven or eight of them. A goodly percentage of a charge's 400+ pellets will not strike the intended clay targets. In having missed them, the pellets will instead travel some 500–600 ft before coming to rest in a predictable pattern (NSSF 1997).[2] As a consequence of years of range operation, fall zones approximating the shape of a trapezoid, and of the order of 5–10 acres, are formed. At this point, the reader should begin to appreciate that pelleted areas, based on spatial relations alone, may not present as many opportunities for accidental pellet ingestion by upland birds as originally thought. Do we really expect that 1000 upland grit ingesters occupy a 5- or 10-acre area? 100? 50? It would be appropriate to next bring into consideration how likely it is that birds, in an overall sense, might seek out a range's fall zone. To contend with this, there is first the reality that many fall zones are not comprised of upland habitat. Marshy areas and waterbodies may commonly constitute the region where fired

[1] On certain days there could be many more shooters, as when tournaments and other competitions are held.

[2] Of course, where the pellet's flight path is obstructed by trees and other formations, accumulations will characteristically appear at the bases of these.

pellets land and accumulate.[3] Before we do anything more vis-à-vis this assumed environmental contaminant issue that stands to threaten the well-being of a certain class of ecological receptor, we should conduct a much-needed tally. We need to ascertain if there are enough instances of viable upland habitat being the ecological repository of spent shot pellets to constitute a familiar case for the nation, and to warrant formal guidance on the subject (US EPA 2011). Where upland habitat *does* describe the fall zone, we next need to apply our knowledge of grit ingester densities to be followed by our knowledge of the specific habits of these species. It is imperative that the proper focus be maintained when refining our initial thought that rampant bird losses result from the accidental ingestion of spent shot pellets. The goal is not to establish *that* such ingestion events can occur, but rather to ascertain the extent to which these events might occur (i.e., the frequency of such events). If the scope of the possible "problem" isn't that large, the one-time perceived matter falls away. There are then, defensible reasons for endeavoring to be so discriminating. Briefly, recognizing when the scope of a would-be issue is insufficiently great to constitute (or better, "formulate") a problem can (a) spare us from wastefully applying precious resources, and (b) help us circumvent exploring issues when we know *a priori* that we don't have the evaluative tools to get us to an answer.

Density information for grit ingesting species, like that for other birds, is freely available. For the purposes of this discussion, a reasonable figure of one bird/acre is certainly a workable one, given the variability among species with regard to this parameter, and in consideration too of birds' foraging and migratory behaviors. It is worth noting that the EPA reports a highest density for the Northern bobwhite (*Colinus virginianus*), a species we will draw on later in this chapter, of five birds/hectare (or about two birds/acre; US EPA 1993). If our analysis were to end here, we might be thinking about five individuals of each of perhaps three or four bird species occupying a fall zone for at least some of their time. This would not seem to make for a sufficiently robust *problem*, especially as we recall that birds are not confined to the fall zones of trap or skeet ranges, and that human activity (much of it in the form of more than occasional gunfire) might routinely create ample disturbance to keep birds away.

It could be said that there is more that is *not* known about grit ingestion behavior than *is* known. Earlier it was mentioned that it is unclear why the birds ingest grit at all. The unresolved nature of that query is undoubtedly the least of our worries, for the open literature provides only general trends information for the elements of grit-ingestion behavior that so critically factor into the ERA topic we are exploring (Best & Gionfriddo 1991; Gionfriddo & Best 1996, 1999;

[3] This arrangement, whether fortuitous in nature, or a function of the deliberate design of a range, serves to protect not only upland biota, but humans as well. In the latter case, other than illegal trespassing, it's unclear why people would have need or want to walk through a fall zone.

US EPA 2011). How many grit particles a given bird species carries about in its gizzard at any one time is not very well known. The same is true for the retention times of grit particles. (It is thought that true grit particles become worn down from the grinding activities, and are either voluntarily or involuntarily excreted after a time.) How often a new grit particle is ingested – to some degree a function of the rate at which grit particles are excreted – is not well understood. Preferences with regard to the color, size, and shape (angularity) of grit particles can strongly influence grit selection, and this only begs more the question of how does any of the information we might possess about the ingestion of *true* grit relate to the accidental ingestion of spent lead shot pellets that deviate substantially from the former.[4] If the reader hasn't come to appreciate this last point yet, its overarching nature might unequivocally suggest that we should not be pursuing the possibility of birds being lead-poisoned at the fall zones of trap and skeet ranges. It is interesting to note that in an attempt to arrive at a better understanding of the likelihood of shot pellet ingestion, some researchers were astute enough to factor into their analysis, the distinct spherical nature of pellets that sets them apart from true grit (Peddicord & LaKind 2000).

To regroup, at this early stage of review we may have already learned that we are not at all on a sound footing when we consider taking up the case of the potential for lead poisoning of birds at trap and skeet ranges. The phenomenon may not be as widespread as we believe. The US might contain some 9000 non-military shooting ranges, but the exposure pathway may not always, or often, be operative at them. The substrate might not be conducive to the behavior and if it were, at any given range (or more correctly at any given fall zone), we might not have enough birds engaging in accidental pellet-ingesting behavior to warrant an assessment. Truly, we should be accountable for knowing over a breeding season or over the course of calendar year, approximately how many birds are likely to ingest a spent shot pellet from a given range's fall zone. Uncertainty will undeniably cloud such a calculation, but the unlikelihood of such an event must be appreciated. A bird might only have five or ten grit ingestion events in a year, recalling that such events are probably triggered in large part by a bird's sensing, perhaps through a sphincter-control system (McCann 1939) that it has excreted a particle that contributed to its species-specific "standing count" (US EPA 2011). Also to be considered is that grit is not essential to sustain the life of a bird. The literature is replete with examples of variable grit types having been set out for birds to take, and where birds did not select any if their preferred kind was

[4] Unlike true grit, lead shot pellets are spherical or nearly so, will register a metallic taste in the mouth, and are considerably denser than equivalently sized bits of rock or stone. Conceivably, birds in the wild might reject shot pellets shortly after ingestion because of these differences. Pellets may not lodge as well against the gizzard's lining, explaining the relatively shorter retention times of shot (versus grit) described in the most current literature.

not among those offered (Best & Gionfriddo 1991). With our knowledge of bird densities, the relatively small size of fall zones, the non-necessity of grit in the diet, the seasonably variable diet of birds (reflecting lesser and greater needs for grit), and the vast uncertainty associated with shot pellets serving as surrogates of true grit, we are really in no position to arrive at the problem formulation stage. With all that transpires ecologically over a breeding season or over a year, and the irregularity of grit ingestion events occurring, we need to consider what the chances are that of all places, a bird will come to have as the site of a particle (pellet) ingestion event, the fall zone of a shooting range.

The supreme point to appreciate is that being wishy-washy with regard to the study question – clearly the case here – does not bode well for the later stages of an investigation. We have egg on our face if we conclude that there is a high probability that a bird will ingest a pellet or will die from ingesting a pellet, if there's only a handful of birds to whom such events are occurring.

The opening paragraphs of this chapter meant to convey that we stand to dig a bigger hole for ourselves when we pursue a question (i.e., a problem formulation statement) that lacks substance. In traditional ERA, the presence of a chemical, by itself, should not trigger an investigation of ecological effects (Tannenbaum 2005). With so few instances *of bona fide* impacts (to terrestrial vertebrate wildlife, in particular) known to us at contaminated sites, we would do best to only peer in occasionally from the sidelines than to throw our mediocre assessment scheme at what we thought to be a problem. The parallel of conventional Superfund-type sites and shot pellet sites is keen. At the former, given the mounds of time that have transpired, we have every reasonable right to know of demonstrated impacts. We don't know of any, but fueled by our bias that there must be an ecological toll taken when chemicals are present, we insist on plodding along, using inadequate tools to try to demonstrate that ecological receptors are paying a heavy price. To date, no one has documented the case of a decimated bird population at a trap and skeet range, nor has anyone ever alleged an approximate number of birds that a particular range is responsible for having killed off. The parallel continues; just as we utilize toxicological data in Superfund investigations that bear little relevance to the chemical exposures that occur in the field, the shot pellet exposure studies that are commonly cited for this specialized type of assessment do not reflect the field condition. Inappropriate studies are of several kinds. Some involve the use of a substantially larger shot size than is used at trap and skeet ranges (Fimreite 1964; Gjerstad & Hanssen 1984), while others force-fed shot at rates that well exceed what is reasonable to expect an upland bird to ingest (McConnell 1968; Buerger et al. 1986; Castrale & Oster 1993; Patee et al. 2006). Studies that replenished force-fed shot each time a pellet was excreted (Patee et al. 2006) constitute another unsuitable type. Such may have utility in another area of toxicology (i.e., where the intent is to document the health effects that occur when there is a continual lead source in the body), but are inappropriate for our needs. Studies that do not report pellet retention times, manifested blood lead levels, standard blood parameters (e.g., red blood cell count, packed cell volume) or delta aminolevulinic acid dehydratase level (a prime indicator of lead poisoning) have limited utility for the subject matter with which we are dealing. In a very real sense, for virtually all the force-feeding lead pellet studies we may be able to salvage for use because they do not share the pitfalls described here, there is still an overarching limitation. The studies will have employed brand new (i.e., unused) shot rather than the spent shot that typifies a fall zone. The latter, lingering for decades on the ground as it weathers, most recently earned the moniker of "environmental shot." In place of having the shiny grey-black metallic luster of unused pellets, environmental shot pellets develop a crust of white or brown material (Jorgensen & Willems 1987) from the formation of lead oxides, carbonates, and other soluble lead compounds brought on by weathering (Sever 1993). There is every reason to suspect that new and environmental shot behave differently inside an upland bird. With a corrosion "cap" formed on environmental shot, pellet dissolution might be comparatively slowed. If pellet retention time is relatively brief, a delay in lead effacing from ingested pellets might mean the difference between a bird succumbing to lead toxicosis or surviving. It should be noted that brand new shot held in the palm of one's hand for just a minute or so, will rather immediately leave behind some of the lead of its exterior, in tell-tale grey-black markings. Such does not occur with environmental shot.

We have already brought forward quite a few pieces of information contraindicating ERA investigation for our case study. The problem formulation falls flat because of the extreme uncertainty associated with likening the phenomena of particle ingestion behavior and particle processing behavior for grit and shot pellets. Problem formulation also falls flat because we know from the start that we haven't an assessment scheme to inform us that a bird is consuming too many pellets. There is more; the case study has an additional complicating problem formulation element that does not occur with conventional Superfund-type sites. Apparently, in the rush (or perhaps, the delight) to hear of a new and specialized type of assessment to embrace, ERA practitioners have set off, half-cocked. Misinformation about the existence of an assessment scheme has drawn in an audience that is bent on repairing the field condition by strategically reducing the number of pellets lying on the ground, this when they are not technically empowered to do so.

The earlier mentioned Peddicord and LaKind publication (2000) has done more to engage ERA practitioners in the case study topic than any other contribution to the field. A careful read of this pivotal article though, makes it clear that the authors developed their model only to enhance the exposure component of the risk equation (i.e., to be able to speak to the frequency with which birds consume pellets). The model does not generate health risks to birds who might consume shot pellets, as in indicating the likelihood that a bird will develop toxicosis, or in predicting the number of birds (or the percentage of a population) that will die from consumed pellets. Quite to the contrary, the model was designed only to estimate the probability that a bird would ingest one shot pellet in its lifetime. To the extent that the model incorporates a fair share of factors that contribute to that estimation (such as the number of appropriately sized soil particles to be serviceable as grit in a standardized soil sample, and the percentage of all appropriately sized particles in a standardized soil sample that are pellets), the model is a worthwhile contribution to science and to our understanding of an ecological phenomenon. For our ERA purposes though, we haven't gained anything; what can we do with the estimated likelihood of a bird ingesting one shot pellet in its lifetime? The worst thing we could do is misinterpret the model's intent. From this author's perspective, it has been saddening to see how ERA practitioners have come to erroneously fill in the gaps. They have assumed that because the model speaks to the probability of ingesting just one pellet, single pellet ingestion is a lethal activity. We do not know this to be the case, however. Presumably, had the model been designed to report the likelihood of a bird consuming two pellets every 6 months, ERA practitioners would feed off of this information as well.[5] Those who have seized on the model's reporting, equating single ingestion events as lethal ones, have further erred. With a knowledge of a fall zone's present pellet densities and the model's estimates, they have erroneously concluded that they are now in a position to run a back-calculation to determine how many fewer pellets per unit area (e.g., square meter) there need be to ensure a protected population. They are then primed for estimating how extensive a ground-sweeping remedial effort is needed. This account amounts to a classic case of "a little information is dangerous." Presumably for our purposes, the poor science to be applied stems from misinterpretation (or just plain careless reading), undoubtedly spirited on by the enthusiasm displayed for a new ERA issue with which to deal.

Let us recap the parallels for the Superfund program general case and the upland bird/lead pellet ingestion matter, with the express purpose of illustrating all that stands to go wrong when a supposed problem formulation is not authentic.

- We believe a Superfund site is problematic for ecological receptors when we come across chemical concentrations in an environmental medium. We similarly believe that exposed spent shot pellets on the ground surface (and even pellets

[5] There is certainly nothing stopping a person from adjusting the Peddicord and LaKind model such that it reports in ways that might be more meaningful to the user (e.g., the estimated probability of ingesting a pellet for a given season). It is somewhat curious that ERA practitioners haven't come to adapt the model to their liking (while still overlooking that the model, in any form, is not a risk predictor).

that are recessed in the upper inches of the soil column) pose certain lethality to birds. "Belief" needs to be taken to task; believing something doesn't have to mean it is true. We need to know more before we allow the assessment process to advance.

- For ecological receptors in conventional Superfund work, we have no "risk" assessment scheme for chemicals mixed (e.g., dissolved) into a medium that present as a concentration. Similarly, we have no means to assess the ingestion risks of birds where lead particles are the potentially offending agents.
- We proceed to interpret HQs of Superfund assessment scheme output, as risk levels, even though they are not. For lead pellet assessments, we proceed to apply the Peddicord and LaKind model and others like it (e.g., Luttik & deSnoo 2004) thinking that we will learn of the probability of one or more birds dying from pellet ingestion events. These model outputs at best though, are exposure estimates only (i.e., estimates of the likelihood that a bird will ingest a pellet).
- In Superfund ERA, we use the HQ (despite its grave limitations) to calculate safe chemical concentrations for receptors (i.e., those that can remain in place in site media). This is done by back-calculating to a medium concentration that equates with a HQ of 1.0. This procedure assumes a linearity of scale for HQs, although such does not exist. For the pellet ingestion phenomenon, after arbitrarily setting an acceptable mortality rate (as the percentage of the shooting range population to succumb from incidental ingestion events[6]), we also back-calculate. From what we think to be a probability estimate for mortality, we arrive at what we believe to be a pellet density that is safe to be left behind (i.e., that will allow for a given bird population to be sustained). As mentioned earlier, the probability estimates of the models used have nothing to do with toxicity or mortality.
- After so-called remedial concentrations are derived in Superfund ERA work, there is a great reluctance among ERA practitioners to implement the cleanup actions that should translate into a chemically cleaner and thereby, safer site. The possibility of doing more damage to the local ecology by remediating (through imposed habitat disturbance) than by leaving the contamination in place is first discovered at this late stage. Only after (erroneous) safe pellet densities are derived, concerns about the difficulty of maneuvering pellet sweeping equipment through a range's fall zone seem to first magically arise. Again, the worry is that more damage will be done through remedial intervention than by leaving contamination (here, the lead particles) in place.

[6] Curiously, where the Peddicord and LaKind model or others similar to it have been applied, ERA practitioners haven't ever bothered to census the bird populations at trap and skeet ranges. It is likely that true populations aren't found at fall zones to begin with.

Several points are worthy of embellishment for the subject case study:

- Peddicord and LaKind model predictions carry no more validity and generate no more concern when values are relatively and eyebrow-raisingly high. If the model told us that a bird that spends time at a range had an 85% chance of ingesting a pellet, this is not necessarily a bad or alarming thing. The astute ERA practitioner will consider how many birds are the subject of discussion, and more pointedly, he/she will research whether there is a population existing at the range altogether to justify the model having been run in the first place. If for argument's sake, the model, with high accuracy and precision, reported that a bird had a 100% probability of ingesting a pellet in its lifetime, this too should not, at face value, carry with it a negative implication. Is a single pellet ingestion event a bad thing? We will return to this fundamental point shortly.
- It would be unfair to suggest that we should always be distrustful of new calculations and models that are put forth to assist in ecological assessment. More than anything else, overstating such a point might only make known our insecurities and our reluctance to adopt new tools for our craft. Curiously, the bird-pellet ingestion issue was not selected for review in this chapter dealing with problem formulation because the Peddicord and LaKind model has a published fatal flaw – although it does. The word "published" is key here; the model authors assembled the necessary equations to legitimately generate certain estimates of pellet consumption, but as their work was set to print for publication, an unfortunate transcriptional error occurred in the paper's equation 2; a division sign was accidently used in place of the multiplication sign. In conjunction with the preceding point, this paragraph's purpose is to illustrate how the availability of a new and perhaps clever or glitzy tool to assist with an ecological risk concern (namely, that birds could confuse spent shot pellets for true grit) has the capacity to devastatingly mislead us. There might not be enough birds at the fall zone of a shot range to establish a concern, and we might not know enough to say that single pellet ingestion is lethal, but such scenario elements become overlooked when innovative data generation arrives. A new or advanced tool does not necessarily legitimize problem formulation. In the case of Peddicord and LaKind, an ERA reputation-damaging error went unnoticed because of the interest to go out and apply the model. Where the blame lies for the unfortunate error is not something for the reader to decide, nor is it germane to the discussion.[7] We might cringe at the realization that every application of the Peddicord and LaKind model has produced erroneous results, to include two applications that have garnered a fair share of attention and spurred on several other pellet assessments (USFWS 2002a, 2002b), but this too is not our issue with which to deal.

[7] How to go about informing the health risk assessment community of the Peddicord and LaKind model error, discovered 10 years after publication, is not a simple matter. It may be that this chapter's attention to the error constitutes the first formal acknowledgement of it.

- The case of the mourning dove (*Zenaida macroura*) is specifically deserving of our attention because ERA practitioners involved with the matter of incidental shot pellet ingestion by birds, invariably point to this species as one likely to be lead poisoned. We might go so far as to label the mourning dove as the *poster child* of this ERA contaminant uptake pathway of interest. The mourning dove is a grit ingester and occurs in all of the conterminous US states. Although the mourning dove may potentially occur at a great many trap and skeet ranges, the groundwork is not laid for authenticating a problem formulation statement with this species in mind. The mourning dove is the most abundant and widely hunted bird in the US, and it bears a staggering loss statistic over the 38 states where it supports a hefty recreational outlet. The total estimated US annual mourning dove harvest is slightly more than 22,000,000 (Sadler, 1993), and curiously, a recent but no longer used harvest figure (Dolton & Rau, 2006), places the estimate at approximately double this already unfathomable number. As biologists we should marvel at this species' astounding resiliency, that is, its capacity to flourish while having its numbers so substantially reduced year after year through deliberate selective removal. With some US states reporting takes of upwards of one million, we would be disingenuous to not square numbers like these against the risk assessments we conduct. Let us consider a state with a yearly take of 200,000 birds, a staggering toll none-the-less. The number is, of course, a rounded one, and it can certainly accommodate an additional 5 or 10 animals, reasonably the number of mourning doves that could be expected to occasionally occupy a trap and skeet range in that state. Why would it be appropriate to engage in an assessment for lead pellet ingestion for the mourning dove at a given range?

It should be obvious to the reader that for the bird species most implicated in the death-by-incidental pellet-ingestion saga – the mourning dove – we cannot construct a case supporting either the need for assessment or possible site remediation. To begin with, there aren't enough mourning doves utilizing a given range fall zone. We then take note of this species' utter resiliency in the face of severe and recurrent population-leveling events occurring on statewide scales, and on a broader nationwide scale; mourning doves are not disappearing, nor does the species appear on any state or federal protective status list. (If the species had a protected status, we surely wouldn't be recreationally harvesting them to the tune of more than 22,000,000 each year!) Other realities regarding shot fall zone occupancy come to bear as well. The conspicuous absence of mourning doves or any other avian species at trap and skeet ranges has never been described. This, not surprisingly, is in line with the reality that owners of trap and skeet ranges have no need to periodically, if ever, census the species that occupy their land. Thus for any trap and skeet range (fall zone) and for any avian species, we have no baseline population knowledge (i.e., for the period prior to range operation) nor any population statistics since ranges became operational. These last points should prompt us to ask if it would ever be evident to anyone frequenting a trap and skeet range that fewer birds are occupying a range's fall zone than used to be the case. This textual assessment leaves us in a disconcerting state. In recent years we have witnessed the matter of incidental shot pellet ingestion by birds become elevated, and to the point where the US EPA has weighed in with a collective review of methodologies for calculating pellet ingestion probabilities (US EPA 2011). Have we though, enough to say that we have on our hands an ecological risk problem in need of attention, and the need to frame such into a formal problem formulation statement? It appears not. All we really know is that lead shot pellets approximate the size of the true grit that birds consume. It is merely conjecture to suggest that regularly, birds mistakenly consume shot pellets in place of true grit, although it is true that on occasion, carcasses of birds have been discovered with one or more shot pellets in the gizzard. There is conjecture too in so confidently professing the reason for not finding carcasses of poisoned birds, namely that lethality does not develop instantaneously, allowing birds to move beyond the fall zone boundary before dying. Along these lines it is a stretch to say that birds readily accept shot pellets, if such a claim can be traced to certain experimental outcomes where pellet uptake was expressly facilitated. Studies that come to mind here are those where birds were confined to small artificial housing, and where true grit was withheld while supplying shot pellets of a consumable size. We are back to *guilt by association*. Because birds *could* ingest lead shot pellets, we are imagining the worst, namely that severe population decimation may result. With mere handfuls of birds present in fall zones though, severe population decimation is not a relevant consideration, and the few losses a fall zone might actually experience are not enough to propel us into study. We cannot use our *imagining* that a problem exists to parlay our concerns into full-fledged problem formulation.

For any potential ecological health risk matter, a proposal to pursue problem formulation must bear up to its fair share of scrutiny. Where imagination has (nevertheless) largely led environmentalists and ecological risk assessors to address the incidental pellet ingestion of birds, this should not come at the expense of failing to review *real* (as opposed to imaginary) phenomenon-specific information where it exists. In the case of shooting ranges, when it was first recognized that grit ingesting birds could mistakenly consume pellets in place of grit, a real temporal element to acknowledge vis-à-vis the potential for lead-induced toxicological effects to be elicited was born.[8] This recognition milestone occurred

[8] In what might be the most embarrassing of circumstances regarding our issue, decades after it was recognized that birds could incidentally consume shot pellets, the US Fish & Wildlife Service (FWS) purchased a shooting range from the Army and continued to operate it as such for over a decade. FWS then contended that the Army was responsible for the accumulated pellet condition and the potential risk condition posed to birds.

decades ago. Might it not be too late then, to start reviewing the potential for lead poisoning at ranges that have been littered with spent shot pellets for decades? We could adopt an alternative perspective regarding the reality of ranges having allowed for lead pellet ingestion to proceed for so long. Since some ranges have submitted to an assessment scheme of some order, we should expect or insist all others to have similarly followed suit. But they have not done so. It would be too convenient to suggest as the reason for the non-conducted assessments, that ERA practitioners have come to recognize another real piece of information, namely that a means for assessing the potential for bird deaths to occur at pelleted ranges does not exist. We cannot give such credit to ERA practitioners when the available models continue to be applied, with ERA practitioners believing that the models empower them to compute safe pellet densities.

Our consideration of the mourning dove in the pellet ingestion case study brings us to an inadvertent discovery that should be related back to the review of toxicological practices in support of ERA in Chapter 4. The mourning dove is one of the more common avian test species in support of TRV generation. The vast annual harvest numbers for this species though, present a rather colossal difficulty for HQ-based ERAs. With a demonstrated ability to replenish its numbers in the face of huge human-imposed annual population reductions, this species is far too resilient to serve in the capacity of a toxicity test surrogate for other birds in the wild, regardless of the chemical stressor and the form in which it occurs. Although laboratory-based studies have induced toxicological effects in mourning doves, in turn giving rise to NOAELs and LOAELS (Sample et al. 1996), there is no rightful need for such benchmarks when we know that the species can be so extremely tolerant of a supreme stressor in the wild (i.e., the shotgun). The reader should not hear this point as the author confusing the high population growth rate and rapid recovery in the aftermath of imposed population leveling of a species, and the toxicological sensitivities of that species. The confluence of the two features is most deliberate. Each time a mourning dove-based TRV is applied in the estimation of hazard for some other avian species (e.g., robin), a significant additional uncertainty is introduced. The complication for the ERA process at large is that if the mourning dove-based TRVs were withdrawn from use (because of the dove's recognized substantial non-comparability with other species), we would be left with a considerable number of chemicals lacking TRVs.

A generic point developed earlier in this chapter was that available relevant information should always be sought out to help avoid unnecessarily being drawn into the problem formulation phase of an ERA investigation. It is a challenge for many to come to grips with, but there is nothing wrong with determining that what one thought to be an ecological risk issue, is actually not. For the highlighted case, several valuable pieces of information have more than served the purpose of directing attention away from study. The only arguments in favor of investigating the pellet ingestion pathway at trap and skeet ranges were weak ones: pellets being the same size as true grit, and accounts of shot pellets observed in carcass gizzards on some occasions. For the highlighted case, a point also raised was that misinterpreted models have misled biologists and ERA practitioners into assuming that single pellet ingestion is lethal and that model back-calculation can lead to defining *safe* pellet densities (i.e., those for which not too many birds would be expected to succumb to lead poisoning).

Is single pellet ingestion lethal to a bird? A new series of experiments, still of the force-feeding genre (for such is really unavoidable) suggests the answer to this quintessential question is a resounding "no." The simple experimental design replayed several times now (Kerr et al. 2010; Gogal et al. 2012; Tannenbaum *in press*) arrives at the truest answer we are likely to ever have, this through its utilization of the environmental shot discussed previously. It is important to note that aside from the studies cited above, and just one other small-scale study involving only five birds of one sex, and of only 7-day duration (Vyas et al. 2001), spent shot has never been employed in any force-feeding experiment. After the Vyas et al. effort, the next closest experimental attempts at duplicating the field condition only marginally succeed at doing so with their attempts at pellet-weathering. McConnell (1968) endeavored to weather shot by placing unused pellets on the ground for a few weeks before dosing birds. Rocke et al. (1997) retrieved spent shot from sediment cores and set these out before sentinel mallards that had been placed within constructed enclosures. For the newest genre of promising pellet dosing studies – aside from the single pellet gavage feature (or perhaps the gavage of just a very few pellets more), and the collection of the critical physiological measures described earlier (e.g., blood parameters) over a time course of 4 weeks or more – pellet retention and dissolution is monitored with radiography. Combining the information gleaned from two studies conducted with the Northern bobwhite, and one study conducted with the domestic pigeon (*Columba livia* f. *domestica*), the following has been observed.

Northern bobwhite:

- 91% survival of males dosed with one pellet;
- 100% survival of males dosed with two or three pellets;
- 100% survival of females dosed with one pellet;
- 82% survival of females dosed with two or three pellets;
- pellet retention (males and females combined) of 21% after 1 week;
- pellet retention (males and females combined) of 7% after 2 weeks;
- 0 pellet retention at 3 weeks post-gavage;

- three-pellet gavaged males manifested blood lead concentrations of more than 25 times the reported severe clinical poisoning threshold (Franson & Pain 2011); three-pellet gavaged females manifested concentrations of more than six times the threshold;
- birds gavaged with up to three pellets bore no gross clinical signs of disease, nor presented with histological anomalies;
- no statistically significant blood parameter measures occur for one-pellet birds relative to controls.

Domestic pigeon:

- 100% survival of males and females dosed with one, two, or three pellets;
- birds gavaged with up to three pellets bore no gross clinical signs of disease, nor presented with histological anomalies;
- birds gavaged with up to three pellets blood had no statistically significant blood parameter measures relative to controls;
- birds gavaged with up to three pellets showed no liver necrosis, kidney inclusions, inflammation, compromised spermatogenesis, compromised ovarian development, ventriculus ulceration, spleen lymphoid depletion, or spleen hyperplasia.

The above observations would seem to make for quite a compelling case. The misinterpreted Peddicord and LaKind model (2000) continues to have people thinking that single pellet ingestion is lethal. We don't find such to be the case, however, when "ecologically relevant pellet gavage" reigns (i.e., where testing involves administering environmental shot as opposed to unused shot, and where dosing, for all intents and purposes, is reasonable and generally limited to just one pellet). With regard to the reporting of bobwhite survivorship statistics, it is certainly fair to combine one-pellet birds with two- and three-pellet ones. Although upland birds are not documented as having more than one grit ingestion event at a time (and certainly not more than one *pellet* ingestion event at a time), with birds nevertheless surviving the additional pellet loads and tolerating them so well, there is great opportunity to underscore how amiss statements of single-pellet-induced peril actually are. Thus, survivorship for one-pellet females was 100%, and it would have been enough to only report this response. That there was 100% survival in two- and three-pellet females as well, is valuable information; it supplies confidence to the one-pellet finding. Male bobwhite tolerance to lead shot is as good as that of females[9], and 100% survivorship for two- and three-pellet

[9] Male bobwhite clear lead from the body/blood more quickly and efficiently than females, and consequently are less prone to developing illness or succumbing to lead poisoning altogether.

males bears this out in a very utilitarian way, given that one death occurred among one-pellet males (out of a combined group of 11). The reporting of the two- and three-pellet survivorship in this case (although such exposures again, are not realistic) is a matter of "when one has data to use to his/her favor, why not report it?". With two- and three-pellet males having 100% survival, certainly the single one-pellet loss is suspect. The researchers here strongly suspect that the lone bird loss reflects unintentional overbleeding that may have occurred during the study. That is, in addition to the considerable blood lead concentrations that arise from pellet gavage (that require several weeks to clear) constituting a formidable stressor for a small bird, blood draws of 0.5 ml on as many as six separate occasions over the course of a study is a distinct second stressor. It stands to reason that overbleeding on one or more draws best accounts for the slight male survivorship inconsistency.

In the environmental shot testing sequence, the bobwhite was the species first used. This in part, reflected the bobwhite having been a common test species in force-feeding pellet studies in the past. It was the anticipation that the pigeons, secondarily tested with (selected as a surrogate for the mourning dove that is so often alleged to succumb to lead poisoning), would fare at least as well as the bobwhite in environmental shot studies. The predicted pigeon high tolerance to lead (e.g., with one-pellet birds manifesting blood lead concentrations that were 80 times that of controls) was largely based on its greater size (mass) relative to the bobwhite. Importantly, the collective bobwhite and pigeon studies suggest that upland grit-ingesters, over the sizes and weights this grouping can assume, are likely to withstand any incidental pellet ingestion events they may incur.

The manifested blood lead concentrations, particularly in the bobwhite, are enlightening and are deserving of discussion. Birds of both sexes and of each pellet treatment group regularly exceeded the reported severe clinical poisoning threshold of 100 ug/dl although displaying no signs of clinical illness. The import of such discovery in general, and an astounding achieved concentration of more than 2700 ug/dl in three-pellet males in particular, relate well to the topic. With no birds displaying clinical signs of illness, but rather going on to steadily clear out nearly all of the lead over a few ensuing weeks[10], we are again seeing the downside of modeling (referring to the derivation of the poisoning threshold) as it is used to assist ERA. The implication is that before going to the trouble of developing a poisoning threshold, it should first be established that grit ingesting birds that might incidentally ingest a spent shot pellet can regularly succumb. In the broader context of this chapter's discussion, we need to know that we have a real or legitimate problem on our hands before we embark on directed ecological assessment analysis.

[10] It is acknowledged that females require considerably more time than males to assume baseline lead levels.

Given our track record of investing energies to elucidate ecological receptor health concerns and to right the anticipated wrongs occurring in the field, only to find that receptors are not so seriously challenged, we would do well to operate in a different fashion. With the chances being very good that a well-intended suspicion will not pan out, an alternative approach is to first thoroughly investigate suspected health-impacting phenomena being real, and taking all the time that may be needed to accomplish this. Many of the approximately 9000 non-military target shooting ranges in the US have been around a good long while, and decades elapsed before any active efforts were taken to intervene through the ecological assessment route. Assuming that grit ingesting birds continue to populate the range fall zones, we still have time to determine whether or not a problem really exists and to possibly save us from (again) having egg on our face. With regard to the case study, conducting ecologically relevant dosing studies was only one way to validate (or disprove) the lead poisoning concern. There are other approaches, and it behooves us to utilize multiple ones to find that there isn't a problem (if that should be the case), sparing us from conducting multiple assessments that aren't needed.

Another study to conduct would involve trapping birds found at or in close proximity to range fall zones (i.e., those with the best chance of having incidentally ingested shot) and establishing if spent shot pellets are found in the gizzard. Radiography on intact birds could be used, or birds could be euthanized and necropsied. It is recognized that this study type is ripe for the naysayer to weigh in on. If shot were found in the gizzard, he/she could be expected to opine that the pelleted birds would soon die, relishing in having had the good fortune to have stumbled onto birds shortly before they had succumbed. If pellets were not found in the gizzard, the naysayer would undoubtedly opine that the wrong birds had been collected. Assuming potential naysayers could be identified, this last challenge could be circumvented. These individuals could be given the chance to select the field sampling locations they believe to have the so-called best chance of supporting disaffected birds. If specimens were found to be free of pellets in the gizzards, or to be behaving normally despite the presence of gizzard pellets, the onus would be on the naysayer to explain what was observed.

And a third and last approach: along the lines of the query at the end of Chapter 4 asking if contaminated media are surely toxic to receptors, we should investigate if grit ingesters actually ingest shot, or if they land on the ground sufficiently at fall zones to do so. What is to stop us from strategically placing surveillance cameras outdoors, mounting them to trees and pointing them at the ground at highly pelleted areas? We can design a motion-detection-activated camera system that records, as vital events, birds landing on the ground where we want to observe the all-important behavior. The reader should be enthused to know that a test system is currently in place to ascertain the viability of this tracking scheme. If at a given highly pelleted area that supplies appreciable habitat, very few birds are found to alight on the ground over many months, how should we imagine naysayers would weigh in this time? Open-minded readers, it is hoped, would come to the conclusion that the incidental pellet ingestion pathway isn't operative, or operative to a degree that legitimizes ecological assessment.

References

Becker, D.S., Ginn, T.C., & Bilyard, G.R. (2009) Comparisons between sediment bioassays and alterations of benthic macroinvertebrate assemblages at a marine superfund site: Commencement Bay, Washington. *Environmental Toxicology and Chemistry* 9:669–685.

Best, L.B. & Gionfriddo, J.P. (1991) Characterization of grit use by cornfield birds. *Wilson Bulletin* 103:68–82.

Buerger, T.T., Mirarchi, R.E., & Lisano, M.E. (1986) Effects of lead shot ingestion on captive mourning dove survival and reproduction. *Journal of Wildlife Management* 50:1–8.

Burt, W.H. & Grossenheider, R.P. (1980) *A Field Guide to Mammals*. Houghton Mifflin, Boston, MA, USA.

Castrale, J.S. & Oster, M. (1993) Lead and δ-aminolevulinic acid dehydratase in the blood of mourning doves with lead shot. *Proceedings of the Indiana Academy of Science* 102: 265–272.

Chapman, J.A. & Feldhamer, G.A. (eds.) (1992) *Wild Mammals of North America: Biology, Management, and Economics*. Johns Hopkins University, Baltimore, MD, USA.

CH2M Hill (2001) *Development of Terrestrial Exposure and Bioaccumulation Information for the Army Risk Assessment Modeling System (ARAMS)*. CH2M, Sacramento, CA, USA.

Dolton, D. D. & Rau, R. D. (2006) Mourning dove population status, 2006. US Fish and Wildlife Service, Laurel, MD, USA.

Fimreite, N. (1964) Effects of lead shot ingestion in willow grouse. *Bulletin of Environmental Contamination Toxicology* 33:121–126.

Franson, J.C. & Pain, D. (2011) Lead in birds. In: Beyer WN, Meador, JP (eds) *Environmental Contaminants in Biota: Interpreting Tissue Concentrations*, 2nd edn. pp 563–594. Taylor & Francis, Boca Raton, FL, USA.

Gionfriddo, J.P. & Best, L.B. (1996) Grit-use patterns in North American birds: the influence of diet, body size, and gender. *Wilson Bulletin* 108:685–696.

Gionfriddo, J.P. & Best, L.B. (1999) Chapter 3: Grit use by birds, a review. In Nolan, V. Jr., (ed) *Current Ornithology*, vol. 15, pp 89–148, Plenum Publishers, New York, USA.

Gjerstad, K.O. & Hanssen, I. (1984) Experimental lead poisoning in willow ptarmigan. *Journal of Wildlife Management* 48:1018–1022.

Gogal, R., Holladay, J.P., Nisanian, M. et al. (2012) Dosing of adult pigeons with as little as one #9 lead pellet caused severe delta-ALAD depression, suggesting potential adverse effects in wild populations. *Ecotoxicology* 21:2331–2337.

Hansen, J.A., Audety, D., Sopears, B.L. et al. (2011) Lead exposure and poisoning of songbirds using the Coeur d'Alene River Basin, Idaho, USA. *Integrated Environmental Assessment and Management* 7:587–595.

Jorgensen, S.S. & Willems, M. (1987) The fate of lead in soils: The transformation of lead pellets in shooting-range soils. *Ambio* 16:11–15.

Kerr, R., Holladay, S., Jarrett, T. et al. (2010) Lead pellet retention time and associated toxicity in Northern bobwhite quail (*Colinus virginianus*). *Environmental Toxicology and Chemistry* 29:2869–2874.

Loss, S.R., Will, T., Marra, P.P. (2012) Direct human-caused mortality of birds: improving quantification of magnitude and assessment of population impact. *Frontiers in Ecology and the Environment* 10:357–364.

Luttik, R. & deSnoo, G.R. (2004) Characterization of grit in arable birds to improve pesticide risk assessment. *Ecotoxicology and Environmental Safety* 57:319–329.

McCann, L.J. (1939) Studies of the grit requirements of certain upland game birds. *Journal of Wildlife Management* 3:31–41

McConnell, C.A. (1968) Experimental lead poisoning of bobwhite quail and mourning doves. *Proceedings of the Southeast Association of Game and Fish Commissioners* 21:208–219.

NSSF (1997) *Environmental Aspects of Construction and Management of Outdoor Shooting Ranges*. National Shooting Sports Foundation, Newtown, CT.

Patee, O.H., Carpenter, J.W., Fritts, S.H. et al. (2006) Lead poisoning in captive Andean condors (*Vulture gryphus*). *Journal of Wildlife Diseases* 42:772–779.

Peddicord R.K. & LaKind, J.S. (2000) Ecological and human health risks at an outdoore firing range. *Environmental Toxicology and Chemistry* 19:2602–2613.

Rocke, T.E., Brand, C.J., & Mensik, J.G. (1997) Site-specific lead exposure from lead pellet ingestion in sentinel mallards. *Journal of Wildlife Management* 61:228–234.

Sample, B.E., Opresko, D.M., Suter, G.W. (1996) Toxicological benchmarks for wildlife: 1996 Revision. ES/ER/TM-86/R3. Oak Ridge National Laboratory, Oak Ridge, TN, USA.

Sample, B.S., Hansen, J.A., Dailey, A., Duncan, B. (2011) Assessment of risks to ground-feeding songbirds from lead in the Coeur d'Alene Basin, Idaho, USA. *Integrated Environmental Assessment and Management* 7:596–611.

Sadler, K.C. (1993) Mourning dove harvest. In: Baskett, T. S., Sayre, M. W., Tomlinson, R.E., & Mirarchi, R. E. (eds) *Ecology and Management of the Mourning Dove*, pp. 449–558. Stackpole Books, Harrisburg, PA, USA.

Sever, C. (1993) Lead and outdoor ranges. National Shooting Range Symposium Proceedings, Salt Lake City, UT.

Tannenbaum, L.V. (2005) A critical assessment of the ecological risk assessment process: A review of misapplied concepts. *Integrated Environmental Assessment and Management* 1:66–72.

Tannenbaum, L.V. (in press) Evidence of high tolerance to ecologically relevant lead shot pellet exposures by an upland bird. *Human and Ecological Risk Assessment*.

US EPA (1993) Wildlife Exposure Factors Handbook, Volume I. EPA/600/R-93/187a. US Environmental Protection Agency.

US EPA (1997) Ecological Risk Assessment Guidance for Superfund: Process for Designing and Conducting Ecological Risk Assessments, Interim Final. EPA/540-R-97–006. US Environmental Protection Agency.

US EPA (2001) Best Management Practices for Lead at Outdoor Shooting Ranges. EPA-902-B-01-001. US Environmental Protection Agency.

US EPA (2011) Assessment of Methods for Estimating Risk to Birds from Ingestion of Contaminated Grit Particles. EPA/600/R-11/023 ERASC-016F. US Environmental Protection Agency.

US FWS (2002a) Ecological Risk Assessment for the Prime Hook Lead Pellet Site, Prime Hook National Wildlife Refuge, Milton, DE. US Fish and Wildlife Service.

US FWS (2002b) Final Baseline Ecological Risk Assessment for the Former Skeet and Trap Range, US Army Transportation Center, Fort Eustis, VA. VA Field Office. US Fish and Wildlife Service.

Vyas, N.B., Spann, J.W., & Heinz, G.H. (2001) Lead shot toxicity to passerines. *Environmental Pollution* 111:135–138.

7 Getting beyond ERA

Knowledge of the world around us continues to accumulate in exponential fashion and at an astounding pace, and newly acquired knowledge invariably leads us to improved ways of doing things. Perhaps this is most true for the realm of science and its various sub-disciplines. We would be hard-pressed, for example, to find an aspect of medicine that hasn't profited from the information we have gleaned through a process of deliberate study and the deliberate utilization of that information. The ongoing stream of improvements that we regularly witness reflects the view that the capability resides within us to introduce changes that are truly needed and that will hopefully be welcomed. It is unfortunate, but unlike other sciences and sub-disciplines, the ERA process is all but static. This would not constitute a problem if the process was perfect or nearly so, but this is not the case. The earlier chapters made it their purpose to point out the process's shortcomings, to illustrate that there are alternative understandings of ERA concepts, and to allude to alternative assessment approaches to be reckoned with. The previous chapter even went so far as to suggest that we might not have ecological health issues at contaminated sites to resolve in the first place! That we have yet to see any attention to this last notion may speak reams about our great preference for applying the one process we know and our indifference to the consideration of alternatives. The present chapter's purpose is to provide an analytical treatment of the prospects for advancing the science of ecological assessment (but not necessarily "ERA" as we know it). Although the prospects for getting beyond ERA are influenced by economics and political agendas, they are undoubtedly too dependent on how ERA practitioners, and especially the movers and shakers in the field, think about the science involved. This latter consideration places us at a disadvantage for we cannot easily peer into the psyche of the ERA practitioner. The author is not being facetious when he suggests that hypnotizing those ERA practitioners that drive the process, to understand where the impediments to the development of better ecological assessment schemes lie, would best serve this chapter's needs. Such doings though, are only the things of which dreams are made. How helpful it would be to hear the response of an entranced upper echelon EPA science-policy expert, when asked if the means exist to express the

Alternative Ecological Risk Assessment: An Innovative Approach to Understanding Ecological Assessments for Contaminated Sites, First Edition. Lawrence V. Tannenbaum.

probability of an ecological receptor developing a toxicological endpoint. In lieu of hypnosis, and until data are available from ongoing studies that utilize other means to expose the thinking of ERA practitioners, a common sense step-wise argument is put forth. Every effort has been made to support the analysis by drawing on actual events and experiences.

Before embarking on the analysis, the reader is reminded of a cautionary note first expressed at the end of Chapter 1. For science to proceed, as it is expected to, it must be willing to accept criticism. Science must be willing to hear that it could be doing a better job, or that an issue at hand hasn't yet been sufficiently tackled. Where truthful criticism cannot be openly received, those on whom the criticism falls are in no position to argue the point. For the case at hand – endeavoring to assess ecological receptor risk for contaminated site settings – the current ERA process does a very meager job and the earlier chapters have appropriated sufficient text to make this clear. We cannot expect the agency primarily responsible for developing ERA guidance and for administering the process as we know it, to enthusiastically receive such negativity. Human nature being what it is, electing to respond will invariably bring forward a defensive posture and perhaps too, character assault directed at the one who expressed the criticism. This book does not have as its purpose the elicitation of a debate over the merits of ERA prowess. Its purpose, especially at this late juncture, is to support the case that alternatives to ineffective approaches to ecological assessment need to be seriously considered. For reasons as obvious as the inability to be objective, it is the author's recommendation that employees of the EPA do not serve as book reviewers.

Is the ERA process in need of change? The question can be transformed to "Is utilitarian information generated under the present scheme?" If we can be honest with ourselves, we will answer the transformed question with a resounding "no". For terrestrial systems, the best we can do is estimate if site receptors are consuming more of a chemical than they can tolerate, but this procedure is highly imprecise. The system we employ regularly tells us that even at pristine environments, receptors are consuming too much of a chemical load and commonly the estimates produced are toxicologically impossible. Generated estimates are frequently for animals that are known to only minimally contact sites (e.g., a red-tailed hawk at a half-acre site), and in an effort salvage the estimates (i.e., to supply some element of credibility), adjustments are made for site area use and temporal site use. To what should be our dismay, benchmarks for plant protection are screened against soil contaminant concentrations at fully vegetated properties. Adding insult to injury, the benchmarks are often based on seedling growth *in solution* (and not soil), and where the seedling test species often include bush beans, barley, and alfalfa, species most unlikely to occur at contaminated sites (Efroymson et al. 1997). For aquatic systems, contaminant concentrations in sediments are often screened against a benchmark set that is best described as a hodge-podge of low-effect or higher effect levels (Long & Morgan 1990); woven together in the set are outcomes of both laboratory and field studies, and outcomes of both acute and chronic toxicity tests for species of multiple phyla for freshwater, brackish, and marine conditions. Exceeded benchmarks are interpreted to mean that sediment remediation is needed. Notably, remedial action can proceed without establishing that benthic communities have shifted, or that shifted benthic communities translate into compromised health effects in higher trophic levels. With only this partial accounting of the minimalist efforts we expend, and the vast uncertainty that we're left with in the wake of these efforts, we have no right to be content.

This leads to the question of why we haven't seen an expansion of capabilities to better understand how contaminated Superfund-type sites function ecologically. Admittedly there are greater environmental problems in the world to address and try to resolve (e.g., global climate change, deforestation, acid rain, and food distribution) than determining if risk is acceptable or not at a Superfund site, and these matters are truly more deserving of our resources. Short of supplying enhancements to present-day ERA though, we are left with an incomplete assessment scheme. We are capable (only) of suggesting that there *is* risk, but we are not capable of indicating how much risk there is – a critical distinction. Importantly, there is no genius involved in stating that there is risk at a site. We can make such a pronouncement armed only with the knowledge that contamination exists. Surely just acknowledging that contamination is present is not enough to weigh in on the risk question. Why else have numerous agencies developed risk assessment programs and associated guidance to qualify and quantify the manifestation of contaminant exposures? There is risk associated with air travel, for airliners do crash. Simply because crashes occur though, we do not conclude that airliners all need to be redesigned, or that people should all switch to other modes of travel to reach distant destinations.

That the ERA process hasn't notably advanced over the years may not be apparent to everyone. It could be that a failure to appreciate the lack of advancement reflects a delusional perception that the ERA process is, quite to the contrary, a dynamic one. It could be that such a perception is fed by the awareness that procedural changes have indeed come about over the years. Despite the changes, some of which are reviewed here, at the end of the day we should observe that the ERA process is most definitely static.

- Over the years, it was noticed that soil-based HQs for aluminum and iron commonly exceeded the threshold value of 1. When it was recognized that these common elements in the Earth's crust couldn't always be the problem-posers, the EPA adjusted their soil screening policy (US EPA 2008), identifying certain soil conditions that contraindicate HQ computation (US EPA 2003a, 2003b).

Specifically, if the soil pH is less than 5.5 in the case of aluminum, or less than 5.0 in the case of iron, direct contact toxicity (for plants and inverts) is generally not a concern, regardless of actual metal concentrations. Although this change reveals an awareness of an aspect of HQ application that needed adjustment, it would not be correct to say that the ERA process has moved on to new ground because of it. For starters, one should consider the minority of instances when aluminum and iron are carried through ERAs as contaminants of potential concern, and especially when iron is a recognized essential nutrient for all living forms. ERA science is not stronger simply because a formal memorandum may have issued from say, the EPA's Office of Solid Waste and Emergency Response to the effect that HQ computations may be dispensed with for two low-profile inorganic species. The larger point to discover here is that as with the aluminum/iron-soil pH issue, most if not all changes that ERA has seen, relate only to HQ computation. We are not getting beyond ERA if we continue to implement the same essential mathematical operation, albeit with some minor tweaking.

- Early on in the history of HQ computation it was recognized that invariably, the weights of the test species used in TRV generation and those of the field species to be assessed differ, at times by a considerable degree. With attention to this, EPA recommended a process of adjusting TRVs through allometric scaling (US EPA 1993). In more recent years, however, the decision was made to abort the scaling practice because the science did not well support the adjustments (Sample & Arenal 1999). Even if dispensing with allometric scaling resulted in truer HQ values being produced, we have seen no betterment in our overall assessment scheme, for HQs are not the be all and end all of ecological assessment. The HQ does not technically empower us to decide if an ecological receptor is at risk or that a remedial action is needed.
- Standard practice in contaminant screening in support of ERA (as with HHRA) used to involve the early comparison of onsite chemical concentrations with those found in background. After a frequency of detection screen, background comparison was a mainstay, logically placed next procedural step. Regardless of chemical category, finding site and background concentrations on par with one another informs that what one may have suspected to be a site stressor is not really so. Going back one decade saw the EPA institute a regressive maneuver – site/background comparisons would be deferred until after HQs were computed (US EPA 2001, 2002). It is critical to appreciate the import of such deferred comparisons. When previously the recommended approach to health risk assessment was one of seeking out opportunities to streamline and simplify (by eliminating chemicals from further consideration when it was clear they didn't belong in an assessment), the move to defer background screening

to much later in the process sends a diametrically opposed message. It is saying that regulators are uncomfortable with moving in a direction that could portray a site to not be so bad after all. By deferring background screening, the door is left open to at least initially show that a site is problematic, this because of the veritable certainty that there will be one or more HQs above 1.

- Attentive to the realities of our toxicity databases not being robust and our inability to pinpoint where safe doses (NOAELs) transition to effect-level ones (LOAELs), the EPA in recent years promoted its less than formalized "Rule of Five", billed as a novel weight of evidence approach to derive a site-specific cleanup goal. Acknowledging the uncertainties associated with HQ estimates (something which we should probably be so pleased to hear), there was recognition of a potential to require cleanups that were more stringent than necessary. With the Rule of Five, the NOAEL–LOAEL range is divided into several nodes (generally five) to facilitate a statistical review of dose–response. Well-intended as the "rule" may be, it nevertheless completely overlooks that the HQ is not a starting point for determining that a cleanup is necessary. As has been mentioned times over, with HQs greater than 1 we do not know that an ecological receptor is in a health-compromised state. For all of the investment of time and energy to craft the Rule, its application will always be premature. Matters are only made worse after a node is selected from a point along the NOAEL–LOAEL spectrum. Completely forgotten is that HQs are not linearly scaled, and thus the procedural back-calculation to a HQ of 1 triggered by node selection amounts to an erroneous computation.
- Calculating a reliable mean chemical concentration for an affected medium to be evaluated in a risk assessment might not always be a straightforward affair. Thus the EPA's early attention to the derivation of the concentration term (US EPA 1992) gave way, beginning a decade later, to an expanded statistical review to produce still better approximations of the exposure point concentration term (Singh et al. 2002; Singh 2004). Computation of the 95% upper confidence limit of a population's arithmetic mean now cautiously incorporates a dataset's specific distribution type (as normal, lognormal, gamma, or other) as it probably always should have. As with the earlier examples, however, the science of ERA is not improved in the least if, in ERA, the HQ is the goal of our efforts. Once again, in the guise of having produced a more accurate and buffered central tendency term, we might come to believe that ERA is always progressing when it is not.

The static nature of ERA is evident too in the manner and pace at which interagency discussions proceed. The following account of an actual stakeholder meeting that convened to discuss a draft ERA illustrate this. With adjustments made only for the names of the agencies involved, and for the physical location of the site in question, the account is one that is replayed on countless occasions still in the present day.

The state regulator at the meeting was cognizant that HQs for the Louisiana black bear exceeded 1. He was aware too that there had been but one anecdotal account of just one Louisiana black bear sighting at the contaminated property, this from some 10 years prior. In the dialogue at the meeting, the responsible party's risk assessment team indicated that the bear issue wasn't alarming, and reminded the attendees that the bear had been included in the assessment only at the insistence of the regulators. This prompted the state regulator, himself not an eco risk assessor, to respond, directing his words at the responsible party's risk assessors: "We can't close out this site until you can prove to me that the site is safe for the Louisiana black bear." The open-minded reader would have to agree that the above account is nothing less than disgraceful. It is so not only because the science doesn't stand up (e.g., we can't be sure that a bear utilizes the site), but because of the unreasonable demand that arose through the science being so lacking. The authoritarian figure being oblivious to the fact that his demand was so unrealistic (i.e., his thinking that a means to provide a proof of an animal's safety conveniently exists) only exacerbates the point; no one could possibly supply the information to quell the regulator's need (that was expressed more as an ultimatum than as a request for more information) and no one should have to. For our needs, if the above scenario typifies how so-called negotiations are still proceeding today, we have perhaps the clearest proof that ERA is static, and we can understand why it is so. The food-chain model approach virtually guarantees that the HQs it generates will exceed 1. To the extent that responsible parties are overpowered by regulators, and unable to argue their point (that the HQ is a screening tool at best, and not a measure of risk), the regulators have it all; they will see to it that failing HQs are used to indicate that risk is unacceptable, that cleanup numbers must be established, and that cleanups occur. Why would a regulatory agency want to dabble with crafting a different assessment scheme when the one they already have works so much in their favor (i.e., ensuring that HQs fail, which in turn enable regulators to demand cleanups)?

The above establishes that the ERA process is static and it also suggests that the perception might be otherwise, because changes in practice *do* periodically arise. It could be that we are fully cognizant of our process being static and that we believe that it has no bearing on our ability to assess ecological risk. And so the highly pivotal questions so relevant to this discussion are: "Can we in fact, assess ecological risk?" "Why would anyone in our field think that we could?" Where we believe that we have the capability to assess risk, there is no need for entertaining alternatives to what we do, and scientific advancement only takes the form of supplying modifications to what we already have in place. A consideration of the Relative Risk Model (RRM), first introduced in 1997

(Landis & Wiegers 1997, 2005, 2007), may well account for our thinking that we are enabled to determine and quantify ecological risk for the garden variety of Superfund-type investigations that are the subject of this book. The model has been applied many times and has been well-received, and the word "Risk" is central to its name. Just the same, it should be evident from its title that the RRM is limited in the sense that it can only express risk or the potential for it, in a "relative" sense. This is a replay then, of this chapter's earlier-drawn distinction between saying "that there is risk at a site" and saying "the site has a risk level of ____". For all of its utility, it should be obvious how non-germane the RRM is to the day-to-day workings of Superfund-type ERA matters. A description of the model will facilitate this observation. In brief, the RRM considers complex and dynamic ecosystems, such as a marine coastal, fjords, and watershed habitats. A volume of data bearing on sources of contaminants, stressors, and disturbances over a regional scale is assembled and consolidated in a GIS format. A map is constructed that divides the study area into risk regions, and a conceptual site model is constructed to reveal pathways of potential risk. A ranking scheme based on the assembled data is then established, permitting so-called relative risks to be calculated by combining rank scores and by screening the scores using exposure and effect filters. At the end of the modeling, the RRM touts the ability to identify the ecological resources at highest risk of being impacted along with their geographic locations, the stressor contributing the highest risk, and the impact that poses the highest risk to the valued ecological resources.

Although conventional Superfund work does not take us to fjords, watersheds and the like, and habitats that span 40 or 50 river miles or more, the above RRM review should first secure the point that the model hasn't the capability to assess ecological risk, i.e., the probability of a receptor developing a toxicological endpoint. In support of RRM, many years' worth of data are collected for descriptors that for all intents and purposes, are never considered at conventional sites – measures or estimates of soil erosion, streambank development, altered surface water flow rates, and channelization. Even with this vastly greater volume of assembled information, the RRM is only capable of ranking sectors of a river reach (i.e., to say that Sector C has greater risk than does Sector E). Importantly, and to build on this example, the RRM cannot inform that receptors at Sector C are at an unacceptable risk level; we only know (or maybe, *suspect*) that Sector C receptors *have it worse* than do Sector E receptors or receptors at some other point in the river. What if receptors that *have it worse* because of the mix of stressors with which they must contend, are functioning well nevertheless? The RRM doesn't appear to be sensitive to this. Before singing the praises of the RRM we should ask what great utility lies in supposedly identifying the ecological resources at highest risk of being impacted. Herein is an opportunity to recall the temporal concerns profiled in the opening chapters. Prior to applying the RRM at a complex and dynamic ecosystem as in Landis and Thomas (2009), soil erosion, streambank development, altered flow rates, and channelization will have categorically occurred for decades. Consequently it's quite late – in ecological receptor time – to have as a goal, the identification of the ecological resources at highest risk of being impacted. Does the RRM mean to suggest that impacts haven't happened yet?!

The RRM review also provides another service to us, underscoring how non-sophisticated the conventional ERA process is. With the latter we do not aspire to anything more than knowing what chemicals we have in soil, sediment, or surface water, and at what concentrations the chemicals occur. With the fate of sites decided on modeled estimates that consider a singular data type, namely the exposure point concentration (or perhaps the maximum detected concentration), we should find it incredulous that our peers could defend the notion that ERAs express a receptor's likelihood of succumbing to its chemical exposures. In short, if the collection of gobs of information in support of RRM applications don't lead to actual expressions of risk, surely conventional food-chain assessments for the far less dynamic sites we know, cannot either. We have no ability to express ecological risk although we might want to believe that we do.[1]

To see that ERA for contaminated sites is not sophisticated shouldn't have to necessitate the review of an approach like the RRM, which without question, is predicated on highly site-specific layered information. To be fair, in contradistinction to conventional ERAs, the RRM may be applied to arrive at assessments of how ecosystem services are being met, but this should not deflect us from observing the utter simplicity of ERA's in the regular case. Thus, predator–prey interaction describes a central element of ecosystems, but never will we find even remote references to Lotka–Volterra equations and analysis (Lotka 1910) within the pages of a conventional ERA. Realistically, such considerations have no real place in our assessments given the physical scale of the sites with which we deal. While ecosystem services are all the talk at the time of this writing (Wenning & Apitz 2012), we would do well to recognize that for the overwhelming number of sites that submit to the ERA process presently in place, there are no ecosystem services to be concerned about. Thus, the reason that garden variety ERAs lack sophistication is not simply because procedurally we do little else than make a mad dash to crank out HQs and EEQs, never taking the time to describe and analyze the status of ecosystem services. While for vast tracts of forest land, ecosystem services is a relevant concept, and one to key into for the greater elucidation of the (sustainable) ecology of our planet, we must appreciate that there are no

[1] Matters are only made worse for ERA applications overall where we insist that we can assess ecological risk. The author recalls a recent Society of Toxicology and Environmental Chemistry-North America annual meeting wherein a presenter on RRM, when asked if the model assesses risk, responded that it surely does.

relevant ecosystem services to consider at upland sites of 5 acres, that so typify our garden variety ERA work.

Two larger points to be made here shouldn't be missed. First, even if Superfund sites were substantially larger than we know them to be, the approach to the processing of sites adopted over many years precludes the incorporation of population fluctuation considerations or other advanced ecological concepts.[2] Second, it is likely that those who call themselves ecological risk assessors have no familiarity with advanced ecological concepts whatsoever; most ecological risk assessors are not ecologists. Further, their non-familiarity with ecological concepts does not stem from an inattention to these in ERAs, but rather from ecological risk assessors never having studied ecology in depth in their college or graduate work.

Unconventionally large sites whose ERAs garner fair amounts of attention in the peer-reviewed literature present exquisite opportunities to overlook that our science is static. A classic case in point would be the ERA for the Coeur d'Alene (River) Basin (CH2M Hill & URS Corp. 2001), a mining megasite, near the Idaho-Montana border, and a significant component of the larger Bunker Hill Mining and Metallurgical Complex that was added to the NPL in 1983. One of the most productive silver, lead, and zinc mining areas in the US, and nicknamed "The Silver Valley", the Coeur d'Alene mining operation has left a legacy of contamination that extends 166 miles across the state of Idaho and down to the Spokane River in Washington state (NRC 2005). As part of its history, Coeur d'Alene mines produced an estimated 130 million metric tons of ore during their first century of operation alone. Many mine tailings throughout the region were discharged directly to the Coeur d'Alene River and its tributaries until 1968 when the practice was prohibited. That a coordinated effort to tackle an assessment of such sheer magnitude could take hold, is something at which to marvel. With the assessment specifically considering 24 avian species, 18 mammalian species, 4 frog species, several salamanders species, and numerous fish species, in addition to terrestrial plants and soil invertebrates, the reader who is not fluent with the Coeur d'Alene ERA, might imagine that the applied assessment scheme was a necessarily distinctive one. With assessment methods other than HQ modeling unknown to us though, reconsideration on the part of the reader should suggest to him or her that the Coeur d'Alene ERA could not break free of "the mold." This was in fact the case; to what should be our great dismay, this high-profile ERA extravaganza, conformed fully to the only assessment scheme with which we are familiar. It first sported spurious bullet-point conclusions such as:

- Risks to health and survival from at least one metal in at least one area were identified for 21 of 24 avian representative species.

[2] HQs > 1 are mindlessly taken to mean that there is unacceptable risk and that cleanup must proceed, and with no intention to study ecosystem function to possibly understand sites differently.

Adding insult to injury, the ERA failed to note how unrealistic many of its HQs were: "The maximum LOAEL-based HQ for lead was 387 for the spotted sandpiper; for zinc, it was 35 for the sing sparrow." Not surprisingly, the ERA committed what should be understood to be two grievous sins at its terminus. It first spoke of preliminary remediation goal (PRG) development, and the prospect of setting PRGs at levels that would be protective of the most sensitive receptor. It then discussed the intent to set PRGs to background levels in specific instances.

To this point it has been demonstrated that the ERA process is static, and it has been suggested that slight and occasional modifications in practice might lead the ERA practitioner to think the science is advancing. It has been suggested too that our infrequent forays into assessing risk at vast, complex, and dynamic ecosystems, which by no means typify the day-to-day fare of ERA, have the great potential to legitimize our hampered and feeble process. This is especially so when accounts of these forays are high-profiled.

A review of the impediments to getting beyond ERA would be most incomplete without a treatment on practitioner resistance to change. Until now, the unchanging nature of ERA has been set within a context of not (necessarily) seeing a need to expand capabilities, to, so-to-say, grow the science. We should not be so naive as to think that it is because the ERA process works so well, or well enough, that our attention has never really turned to charting a new course. Thus we must be willing to consider that resistance to infiltrating ERA with new ideas, and dramatically revamping the present process, if this is what's needed, traces to a deliberate stance. That the regulating community is resistant to change, or what we might term "ERA reform", is easy enough to demonstrate. After securing this point with two specific factual accounts, an exploration into what might plausibly lay the basis for the regulating community's complacency with the present ERA process will follow.

In 2007, the EPA's Science Advisory Board produced the report "Advice to EPA on Advancing the Science and Application of Ecological Risk Assessment in Environmental Decision Making" (US EPA 2007a). A public teleconference to receive comments on a late-stage draft report version, constituted a unique opportunity to have the final report live up to its name (i.e., to capture concepts that could truly advance the science). One teleconference participant[3] articulated both a series of valid shortcomings of the current ERA paradigm, and an equally valid suggested approach for rendering definitive determinations of the health status of actual site receptors[4] (US EPA 2007b). The teleconference chairperson's rejoinder was eye-opening to say the least (and saddening as well vis-à-vis prospects for moving beyond the ERA paradigm in vogue), remarking that only comments related to refining the present ERA process (US EPA 2007b) were being

[3] this book's author

[4] discussed in Chapter 9.

sought on the teleconference and for the SAB report overall. For those who may have been in attendance on the call, the rejoinder constituted a defining moment for the annals of EPA-led ERA. Call attendees learned right then and there that to the EPA, "advancing the science" only amounts to adhering to the status quo albeit with occasional and minimal modifications made. This sentiment was concretized in the final SAB report (US EPA 2007a) an excerpt of which follows:

> The SAB's Ecological Processes and Effects Committee (EPEC) finds that guidance is needed in the area of risk calculation and application. Too often ecological risk assessments are designed and executed solely with a comparison of measured exposure concentrations to toxicological reference values; the "hazard quotient". Although the area of risk characterization was explored during the EPEC ecological risk assessment workshop, actual discussions around risk calculation and application methods were limited. Much of the workshop discussion in this area focused on the need to better understand the appropriate use of HQs to assess and subsequently manage risk, the issues of uncertainty in calculating risk, and the need for full exploration and disclosure of uncertainty. Therefore, in the following discussion and associated recommendations there is an absence of explicit recommendations related to risk calculation methods other than those related to HQs, which some might suggest are not risk calculation methods at all. The Agency and other readers should not take this as a suggestion that the capacity to do better risk calculations could not be improved. Our expectation and hope is that as EPA seeks to address the recommendations related to appropriate use of HQs and the need to understand and reduce uncertainty, the Agency will need to explore a range of risk calculation methods which represent better and more certain approaches to estimating risk.

EPA is most likely aware that there are some scientists (Tannenbaum 2003, 2005) who challenge the veracity of ecological risk assessments as a tool for effectively informing environmental decisions makers. Although such positions are not widely held, the Agency should not summarily dismiss issues raised by these contrarians. For example, some raise the issue that HQs do not provide estimates of risk (Tannenbaum et al. 2003) and that the use of HQs to set preliminary remediation goals at contaminated sites is mathematically inaccurate. EPEC recommends that EPA take the initiative on this point and develop guidance on the appropriate and acceptable use of such screening tools such as HQs, hazard indices, and other environmental benchmarks, especially with regard to their utility in setting actionable environmental protection goals.

To the open-minded, the excerpt is incriminating for the SAB because of the errors it contains. The statement that "too often ERAs are designed and executed with only a HQ calculation in mind" should have read "ERAs to this point *are only* designed and executed with a HQ calculation in mind." The author is unaware of ERAs for contaminated sites that do anything but produce HQs or EEQs as their terminal step, and he challenges others in the field to furnish examples of assessment goals that are not played off the essential HQ construct. The excerpt mentions "the uncertainties in calculating risk," and this suggests that HQs and the like are expressions of risk, albeit ones that could stand to have some fortification. It is astonishing that the SAB could conceive of HQs as risk expressions when EPA guidance and other formal texts speak to the contrary (US EPA 1989; Kolluru 1996). Where the excerpt's first paragraph alludes to difficulties associated with risk calculation methods in ERA, it alludes to those who "might suggest (that HQs) are not risk calculations at all." Thus, that HQs might not be risk expressions is couched in the context of a debatable point, when in truth, it is an indisputable fact that HQs are nothing more than unitless ratios incapable of expressing probabilities of outcomes. The astute reader should find the excerpt's reference to the allegation that setting preliminary remediation goals (PRGs) based on HQs is mathematically inaccurate, somewhat curious. HQ-based PRGs are unequivocally mathematically inaccurate for several reasons, as earlier discussed. First, deriving PRGs assumes that HQs above 1 signify risk, and we do not know this to be true. Second, as PRGs are chemical-specific derivations (only), there are no provisions for their use with chemical mixtures, which occur at nearly all sites. Finally, dose–response is at best only linear over a portion of the testing range of a chemical. To assume linearity of dose–response for all situations cannot help but lead to mathematical inaccuracy, and especially when there are other chemical stressors present. The reader, or more correctly, the SAB, should contemplate how mathematically accurate the soil PRG for lead for the spotted sandpiper at the Coeur d'Alene River Basin would be if it was derived by dividing the LOAEL-based HQ of 387 by 387.

The SAB's use of *contrarian* is curious too with regard to this chapter's treatment of impediments to introducing improvements to ERA. The implication is that because those who challenge the workings of the current ERA process likely constitute a minority, attention to the points they raise may not always be expected to be heard and/or addressed. That said, there is promise nevertheless in the excerpt: ". . . the Agency should not summarily dismiss issues raised by these contrarians."

The excerpt's parting thought is troubling, and should not be overlooked. While it recommends that EPA take an initiative to do better ERA work (i.e., a promising note it would seem), the recommendation concerns finding ways to use HQs, correctly identified here as screening tools, in the setting of actionable environmental protection goals. How much better the prospects would be for ERA reform if the SAB had the secured understanding that *screening* tools, as their name implies, are not *assessment* tools. This last excerpt recommendation is for all intents and purposes, giving license to mine still new capabilities from the HQ construct. For the purposes of this chapter it is essential to see that fully new approaches to ERA cannot be birthed until we relinquish a hold on the HQ.

With the online availability of the SAB report since its release in 2007 (US EPA 2007b), the reading audience has been sizeable. While projecting the number of readers is not easily done, conceivably all elements of those working in the ERA field were reached. Although garnering the attention of a vastly smaller audience than that of the SAB report's readership, another recent EPA-led venue sent the (apparently) timeless message that the EPA is resistant to considering alternative means for ERA. The setting was the 2010 annual meeting of the Society of Environmental Toxicology and Chemistry's North American (SETAC-NA) chapter, convening in Portland, OR. The specific venue was a half-day platform session entitled "Ecological Risk and Related Ecological Assessments at EPA: Twelve Years after the Guidelines"[5], co-chaired by a top EPA ERA personage. The platform closed with a 20-minute open-mike session: "The Future of ERA: Audience Questions, Answers, and Dialogue", wherein attendees were invited to come forward to bounce ideas off of the panel, and where the panel was to take notes and provide early feedback on what they heard. The session had a facilitator walking through the audience to initially field suggestions before referring to the panel, the topical points raised. For every question, remark, and suggestion except one, the facilitator and the panel engaged in brief receptive dialogue and jotted down information. The exception was for the individual who stepped to the podium and who first noted that ever since ERA came on the scene with the creation of the Superfund program, there has never been a way to express ecological risk. The individual continued, briefly explaining that intending to characterize ecological risk could be a mute point given that decades pass from the time a site becomes contaminated until it submits to investigation. The individual moved on to a third and final point, informing all in the room that a recently patented method, the first ever for ERA, allows for directly assessing the health of ecological receptors in the field at contaminated sites. The individual who came forward, anxious to hear feedback from the facilitator and the panel, found both to be absolutely silent. Further and notably, the key EPA ERA personage bowed his head to stare blankly at the tabletop where he sat. The silence was only broken after several seconds with the facilitator asking if there was anyone else in the room who wished to come up to the podium with a comment or question.

In the best of all worlds, scientists working within their specific disciplines, strive to improve the quality of what they produce. Motivation to bring about substantive changes and advancements can vary substantially. For some, wanting to arrive at some entity's truest measure, or to grasp as absolute, an understanding

[5] Referring to US EPA 1997.

of a phenomenon as the science will allow, provides the push to excel. For others it may be the desire to outdo the competition, and to secure credit for having made the most telling contribution, that drives people to press onward. In the latter case, it would be ideal if scientists did not resort to cutthroat practices so as to keep others from making strides, but we are not so naive as to think that professionalism universally reigns among all colleagues. We have not seen the construction of the very last atomic particle accelerator, or the last experiment designed to learn of the amount of energy that can be liberated when subatomic particles collide. The secrets of the human genome are still being researched, efforts to better predict weather are still being explored, ongoing efforts proceed to identify new species that populate the planet, and we continue to develop reliable and efficient alternatives to fossil fuel combustion to meet the world's energy needs. In contrast, no one, for even a moment, can suggest that we have approximated the zenith of our abilities in the field of ERA. The field has stagnated for more than two decades with the same anemic measure (the HQ) always as its output, and with the powers-that-be referring to it as both a screening tool, and impossibly too at the same time, a risk measure. It is considered a gain for the field when a new animal study offers a TRV, but rather consistently overlooked are the artificial manipulations of the test chemicals prior to their being set before the test species (e.g., pulverizing, heating, mixing into corn oil as a carrier/vehicle). We may want to know how a lizard responds to an ingested explosive compound, but not necessarily when the explosive has been adulterated to an extreme. Lizards in the wild at military testing grounds don't encounter explosives in gently warmed homogenous corn oil suspensions (McFarland et al. 2009, 2011). How often, if ever, is the weight loss associated with a test animal's dosing regimen correctly recognized to be a reflection of the test substance's objectionable taste (Johnson et al. 2005)? Two of three chemical uptake routes are forever ignored, and boilerplate language is habitually used to defend ingestion as the only route of importance. Chemicals are reviewed singly although ecological receptors have integrated chemical exposures. Behavioral and neurotoxic effects are said to be toxicological endpoints of great concern, but there have been no advancements to allow for the assessment of these. And with regard to HQ outcomes, we have no qualms over reporting a LOAEL-based HQ of 387 for spotted sandpipers, or a LOAEL-based HQ of 35 for a song sparrows – as we saw in the highlighted case of the honor-accorded landscape-scale ERA for the Coeur d'Alene River Basin. How is it that spotted sandpipers can survive to perpetuate their own for more than 100 years while daily consuming a chemical at 387 times the effect level for the reproductive endpoint? Doesn't a LOAEL-based HQ of 387 guarantee that spotted sandpipers at the site are severely if not totally reproductively incapacitated?

Why is ERA not advancing? Why is there such (apparent) complacency with the way ERA proceeds (with its imprecision and lacking sophistication)? Why is the regulatory community resistant to change? Plausibly, bias is the great contributor to ERA's stagnant nature. Although not yet proven, it is fair to suggest that the regulatory community cannot come to terms with the notion of leaving known contamination in place in an animal's habitat (or perhaps in any site setting). Going a level deeper, it is fair to suggest that the regulatory community finds it incomprehensible that an animal's habitat can be contaminated, with the contamination not evoking an incapacitating toxicological response. It may be that such sentiments are deeply recessed in the psyche such that they cannot be expressed, but the sentiments nonetheless influence our process. It is one thing to be genuinely concerned about the well-being of ecological receptors that don't have the liberty to move away from the contaminated media they regularly contact. To assume that ecological receptors are regularly suffering chemical impacts, or are unquestionably soon to do so because of their chemical exposures, is something else. To claim that risk is unacceptable when ERA community members would almost assuredly come up dry if challenged to summon forth demonstrated examples of ecological havoc at the sites with which they work, is to display clear scientific bias. A researcher in the laboratory with a decided hunch on how his or her test animals will respond to an administered dose, would not publish findings without first conducting the administered dose experiment – or so we would hope; the scientific community does not furnish proof in the form of stated beliefs. One can only wonder why then, in the aftermath of hundreds and thousands of completed ERAs that did not identify ecological effects, nearly all stakeholders only see the proverbial glass as half-empty. Could it not just be that ecological receptors are far more resilient to contaminant exposures than we realize? Seemingly the ERA community[6] is forgetting that the hundreds and thousands of completed ERAs reflect a great luxury, namely the ability to have examined and reported on receptors that have had dozens of generations of site exposure (Tannenbaum 2003, 2005). Isn't it fair to suggest that over the decades that have transpired, and the dozens of generations that have lapsed, chemicals have become less bioavailable and receptors have had time to successfully adapt?

Plausibly the bias spoken of is but an element of a larger psychological condition known as cognitive dissonance (i.e., the state of psychological conflict or anxiety resulting from a contradiction between a person's simultaneously held beliefs or attitudes). Modern day examples of this condition would include being fully aware of the harmful effects of cigarette smoking, yet having the smoking habit, and vehemently opposing the inhumane slaughter of animals hunted for their pelts, yet purchasing and using fur products made from them. It is entirely possible that ERA for Superfund-type sites is fertile ground for cognitive dissonance to

[6] The ERA community consists of ecological risk assessors, and the great diversity of stakeholders, to include federal and state environmental regulatory bodies, natural resource trustees and their affiliates, Department of Defense agencies, and affected community residents.

be a powerful force at play. Conceivably in this application, members of the ERA community find themselves confronted with two contradictory and simultaneously held beliefs. They believe (and actually do *know*) that *bona fide* contamination is present in the soils, surface water, sediments, and biota of the sites they oversee. They also believe that the healthy ecology described in the reports they read, and the apparent healthy ecology they themselves observe when visiting sites, is real. This arrangement cannot help but give rise to psychological conflict and anxiety. We have all been educated to understand that chemicals are often harmful, and it has been ingrained in us as well since the dawning of the environmental era, that contaminated media stand to wreak havoc with wildlife. It simply must be, our minds tell us, that where ecological receptors are exposed to chemically contaminated media, devastating effects cannot help but take hold. ERA community members could acknowledge that they have their facts wrong, and that they err when they universally apply the "rule" that chemically exposed animals are necessarily harmed. It is suggested here, however, that more often than not, ERA community members resolve their simultaneously held beliefs differently.

Acknowledging that one's personal view or preference is incorrect in the face of more telling information is one of two options for diffusing the anxiety born of cognitive dissonance. (Thus, a smoker could admit that he/she is being hypocritical in keeping to habitual cigarette use.) The second option, and the one more frequently invoked, involves rationalizing the situation such that a person does not find himself to be incorrect. (Thus, a smoker could summon forth one or more reasons why smoking should not put him/her at increased risk of disease, such as smoking fewer cigarettes/day than most, using a filtered brand, or not inhaling the smoke.) Further, electing to invoke this second option reflects man's strong desire to be right, and his strong dislike of being wrong (Robinson & Steinman 1993). To resolve the supposed conflict of expecting there to be a decimated ecology at contaminated sites, and not finding such to be the case, rationalization can assist. Thus, one could argue that the level of effort expended in characterizing a site's terrestrial ecology is insufficient to turn up evidence of grave impact. Alternatively stated, the sentiment is that had more time been spent in the field, and had more measures been collected, chemical-imposed impacts would surely have been noted. Indeed this could be a valid argument, for characterization at prototypical Superfund sites routinely amounts to no more than jotting down observations in a field notebook over a 1–2 day site walk-through. Vegetation is broadly characterized, and based on direct or perhaps mostly indirect evidence (e.g., footprints, scat, sounds/calls, nests, dens, burrows), a list of ecological species present is compiled. The upshot of this analysis is that there is first a (biased) belief within the ERA community that harm is always being done to a site's ecology. Not wanting to be found wrong with that expectation, we seek out ways to demonstrate why health effects are undoubtedly setting in, and why too, the effects might be going on undetected.

Before reviewing a series of plausible bases for the complacency we have with the current ERA process, and why there is resistance to improving the science, it would be useful to return to an earlier point. This analysis has offered that the ERA community expects there to be ecological impacts at sites, a bias outright. It doesn't take much to have such an expectation morph into the wish to find that sites actually have ecological issues that need to be addressed and ameliorated. Demonstrating that we want to have problems to address, or better yet, that we can create problems at will when we wish to, is not difficult to do. Not surprisingly, an over-allegiance (once again) to the HQ construct, albeit from a context removed from ERA, can illustrate the point.

More than 40,000 organic chemicals have been identified as contaminants of emerging concern (CEC), and many of these are poorly characterized in terms of their presence in aquatic environments where they have the potential to pose risk to aquatic wildlife and humans (Diamond et al. 2011a). CECs include pharmaceuticals, personal care products, natural and synthetic hormones, surfactants, current-use pesticides, flame retardants, and plasticizers. All enter the environment through households and wastewater treatment plants that do not completely filter out chemicals. In order to monitor, assess, and help focus the screening of ecological effects due to CECs, there is a definite need for a prioritization system. A catalyst, in part, for endeavoring to assemble working prioritization schemes is the reality that our ability to detect trace levels of certain organic chemicals continues to far outpace our ability to understand how these measurements translate into ecological risks. Efforts to develop working prioritization schemes for CECs for aquatic ecological applications have been particularly challenging, given that different agencies have developed widely variant definitions for the broad category of monitored CECs, this reflecting their individual agendas. In an effort to prevail nevertheless with the daunting task, the Water Environmental Research Foundation (WERF) crafted a framework comprised of three distinct approaches (Diamond et al. 2011b). The first and operationally most simple of these is the hazard-based approach. In essence, a Hazard Value (HV) is computed by establishing the ratio of a CEC's maximum occurrence concentration to its lowest chronic toxicity threshold (i.e., the CEC's most sensitive predicted endpoint, based on either chronic toxicity or estrogenicity effect; US EPA 2011b). The reader should unquestionably appreciate the ultra-conservative nature of the HV ratio (i.e., electing to pair up the highest detected CEC concentration and the lowest toxic effect level, when neither comes close to reflecting the central tendency statistic for the measure and metric, respectively). The denouement to this review is that where an exceedance of an HV of 1.0 was originally used to signify a requirement for ongoing environmental monitoring and ecological management, only some three or four CECs were profiled! Environmentalists should have been overjoyed to hear this; seemingly for at least one of the many global environmental issues we face, things did *not* turn out to be as dire as might have been expected. But we are not done; WERF, for its final report on diagnostic tools for evaluating impacts of trace organic compounds (Diamond et al. 2011b), adjusted the hazard-based approach, setting the HV exceedance threshold to the 10-times more stringent level of 0.1. As a consequence, where once there was no real CEC issue with which to contend, the list of CECs to monitor had, in an instant, ballooned to over 40. The published reason for the threshold resetting was so as to include chemicals that could conceivably present an ecological concern, this while also recognizing the possibility of potential effects being underestimated. This thinking recognizes that underestimates occur either because of the challenge to accurately detect chemicals in surface waters, or because of uncertainties associated with CEC effects that are based on predicted effect values that are used. This published justification appears reasonable, but are we to believe it? A principal scientist responsible for the WERF effort and its reset HV threshold was questioned on the topical matter at a presentation given at the 2011 SETAC-NA meeting. His response made it clear that without the threshold change, a CEC monitoring program might all but fall away. The above then, would seem to make for a rather open-and-shut case of our abilities to create environmental work for ourselves when it isn't needed. It smacks of our uneasiness in reporting that some element of the environment isn't problem-laden. We are now primed to review suggested reasons for our complacency with an ERA process that is so subject to valid criticism.

- *The regulatory community doesn't want to deal with the programmatic consequences associated with acknowledging that sites are not harmful to the ecology.* This assertion posits that the regulatory community sees that the established ERA framework should be restructured, and that formalized ERA proceedings should be dispensed with. The regulating community though, not wanting to shoulder the vast undertaking of revamping programs, rationalizes the situation. Regulators pride themselves in having expended great efforts to establish essential ERA programs, policies, and guidance that should make for a cleaner environment. To rescind program elements that they worked so hard to establish would amount to a (perceived) grave loss. Rationalization also derives from the regulating community's perceived need to appear tough on ERA. Over the years the regulating community has come to be known for routinely introducing additional ERA requirements, and not for lessening the number of them. Although the regulating community well recognizes that they are promoting excessive and unnecessary ERA tasks (because they know deep down inside that there really aren't ecological problems), there is good that proceeds from this. Being aggressive and demanding, insisting that cleanups proceed in order to afford protection to ecological receptors, instills public trust and confidence in the way contaminant issues are managed, and helps to maintain image.

- *ERA community members cannot imagine that they might be wrong in their judgment.* Here, ERA community members rhetorically ask "How is it thinkable that site ecological receptors are not being harmed, when receptors live amid contamination and are often intimately associated with it?" Thus, there is every need for an assessment scheme such as we presently have. ERA community members are as convinced about the certainty of ecological peril occurring at sites as they are about the present process working well. Complacency with the process we have, follows from thinking the process is robust and accurate. Conceivably ERA community members remind themselves that they can point to any number of scientifically defensible laboratory studies where toxicological responses have been elicited in chemically exposed animals. Overlooking the non-comparability of the chemical exposures of laboratory test animals and those of contaminated site animals, it is reasoned that since chemicals had an effect in the laboratory, the same should be occurring in the field. ERA community members may also err with the construction of another well-intended but inappropriate extrapolation. With indisputable evidence that animals in nature, under certain circumstances, have been harmed (e.g., DDT-mediated eggshell thinning in eagles; large-scale wildlife losses due to oil spills at sea), it is taken as a foregone conclusion that ecological harm occurs also at prototypical Superfund-type sites. Here again, the non-comparability of the two cases considered is overlooked. To the suggestion that overt signs of stress or impact are almost completely unknown at Superfund-type sites, errant thinking continues with recalling that at conventional sites, the luxury of being able to defensibly demonstrate that ecological receptors are unharmed does not exist.[7]
- *A cleaner site is a better site (isn't it?).* The sentiment here, independent of ERA community members being capable of or willing to express it, is that it can only help to clean up sites. The mindset of "What better can anyone do than to remove site contamination (i.e., chemicals present in the soil today that simply weren't there originally)?" prevails even in that hypothetical case where it is indisputably demonstrated that soil chemistry poses no ecological concerns whatsoever. With (a) the intent to always clean up sites, (b) the likelihood of computed HQs for site ecological receptors routinely exceeding values of 1.0, and (c) ERA community members having successfully lobbied for decades that

[7] Chapter 9 however, presents information quite to the contrary.

HQs > 1 warrant cleanups, it stands to reason that ERA community members would not want the ERA process changed.

- *There is a grave need to save face.* For ERA community members to now admit that they see no need to continue with ERA investigations at their sites, and that contamination can be left safely in place, is to admit that over the years they have been repeatedly wrong. Further, for community members to now acknowledge that contaminated sites might not be harming the ecology, is to admit that they have been unduly slow to develop the awareness that sites are marked by the conspicuous absence of ecological health impacts. Arguably, complacency with the ERA process is aided by rationalization that takes the form of ERA community members convincing themselves that when Superfund-type programs first came about, assessment practices and skill sets were admittedly weak. Although hundreds if not thousands of time-consuming ecological evaluations and eco-based cleanups have likely proceeded unnecessarily, these actions were done with the best of intentions to abate presumed risks. Community members coming forward today to avow their ineptitude would be damaging to the populace. The public wouldn't take well to hearing that the "experts" erred many times over. To maintain the public's trust in their environmental work, ERA community members have no choice but to maintain the status quo.
- *We are complacent because we are too entrenched in a way of doing things to try another approach.* It is suggested here that the ERA community is cognizant of the standardized approach to ERA being firmly rooted to the point that it cannot be replaced by another approach. This is unfortunate because the standardized approach is costly and fails to yield useful information. Historically expended efforts are viewed as akin to the "sunk costs" of a business venture. With every state and local environmental agency applying the standardized approach or a slight variant of it, there is fear of reprisal should a replacement approach be put forth. Additionally, there is an awareness that no thought has been given to addressing ecological concerns in a different way. We are complacent with the in-place process because we have no alternate process to take its place.
- *We are complacent because we have no available damage control strategy for having overstepped our bounds.* For risk-based programs such as Superfund, the objective is only to ensure that sites do not pose unacceptable risk or hazard (US EPA 1990); there is no requirement to rid a site of its contamination. The ERA community, however, has consistently worked at returning contaminated sites to their pre-release conditions without having demonstrated that impacts are likely to occur or have already occurred. In so doing, the ERA community has often significantly strayed from its legal requirement. Since regulators have

nevertheless "gotten their way" for decades (i.e., cleaned up sites because of chemical presence only), there is no incentive to implement a process other than the in-place one that is mistakenly understood to indicate when ecological receptors are at risk.

To the extent that ERA practitioners can rise above (a) bias, (b) complacency with our present assessment schemes, and (c) an unwillingness to think of contaminated sites differently, the remaining chapters point the way towards alternative – and decidedly improved – ecological assessment.

References

CH2M Hill & URS Corp. (2001) Final Ecological Risk Assessment: Coeur d'Alene Basin Remedial Investigation/Feasibility Study. URS DCN: 4162500.06200.05.a2.CH2M Hill DCN: WKP0041. Prepared for US EPA, Region 10, Seattle, WA, by CH2M Hill. Bellevue, WA, and URS Corp., White Shield, Inc., Seattle, WA. May 18, 2001.

Diamond, J.M., Latimer, H.A., Munkittrick, K.R., Thornton, K.W., Bartell, S.M., & Kidd, K.A. (2011a) Prioritizing contaminants of emerging concern for ecological screening assessments. *Environmental Toxicology and Chemistry* 30:2385–2394.

Diamond, J.M., Thornton, K.W., Munkittrick, K.R., Kidd, K.A., & Bartell, S.M. (2011b) Diagniostic Tools to Evaluate Impacts of Trace Organic Compounds, Final Report. CEC5R08. Water Environment Research Foundation, Alexandria, VA, USA.

Efroymson, R.A., Will, M.E., Suter II, G.W., & Wooten, A.C. (1997) Toxicological Benchmarks for Screening Contaminants of Potential Concern for Effects on Terrestrial Plants, 1997 Revision (prepared for US Department of Energy), Oak Ridge National Laboratory. ES/ER/TM-85/R3.

Johnson, M.S., Gogal, R.M., & Larsen, C.T. (2005) Food avoidance behavior to dietary octahydro-1,3,5–1,3,5,7-tetranitro-1,3,5,7-tetrazocine (HMX) exposure in the northern bobwhite (*Colinus virginianus*). *Journal of Toxicology and Environmental Health* 68:1349–1357.

Kolluru, R. (1996) Health risk assessment: principles and practices. In: Kolluru, R., Bartell, S., Pitblado, R., Stricoff, R.S. (eds.) *Risk Assessment and Management Handbook*. McGraw-Hill, New York, NY, pp.10.3–10.59.

Landis, W.G. & Wiegers, J.A. 1997: Design considerations and suggested approach for regional and comparative ecological risk assessment, *Human Ecology and Risk Assessment* 3:287–297

Landis, W.G. & Wiegers, J.A. (2005) Chapter 2: Introduction to the regional risk assessment using the relative risk model. In: Landis, W.G. (ed). Regional scale ecological risk assessment using the relative risk model. Lewis Publishers, Boca Raton, FL, USA.

Landis, W.G. & Wiegers, J.A. (2007) Ten years of the regional risk model and regional scale ecological risk assessment. Hum. Ecol. Risk Assess. 13:25–38.

Landis, W.G. & Thomas, J.F. (2009) Regional risk assessment as a Part of the Long-Term Receiving Water Study. Integrated Environmental Assessment and Management. 5:234–247.

Long, E.R. & Morgan, L.G. (1990) The Potential for Biological Effects of Sediment-Sorbed Contaminants Tested in the National Status and Trends Program. NOAA Technical Memorandum NOS OMA 52, Seattle, WA, USA.

Lotka, A.J. (1910) Contribution to the theory of periodic reaction. J. Phys. Chem. 14, pp 271–274.

McFarland, C.A., Quinn, M.J., Bazar, M.A., Talent, L.G., & Johnson, M.S. (2009) Toxic effects of oral hexahydro-1,3,5-trinitro-1,3,5-triazine in the western fence lizard (*Sceloporus occidentalis*). *Environmental Toxicology and Chemistry* 28:1043–1050.

McFarland, C.A., Quinn, M.J., Boyce, J. et al. (2011) Toxic effects of oral 2-amino-4,6-dinitrotoluene in the western fence lizard (*Sceloporus occidentalis*). *Environmental Pollution* 159:466–473.

NRC (2005) *Superfund and Mining Megasites: Lessons from the Coeur d'Alene River Basin.* The National Academies Press, Washington, DC.

Robinson, G., Steinman, M. (1993) *The Obvious Proof – A Presentation of the Classical Proof of Universal Design.* CIS Publishers, New York, USA.

Sample, B. E., Arenal, C.A. (1999). Allometric models for interspecies extrapolation of wildlife toxicity data. *Bulletin of Environmental Contamination and Toxicology* 62:653–663.

Singh, A., Singh, A.K., & Iaci, R. (2002). Computation of the Exposure Point Concentration Term Using a Gamma Distribution. EPA/600/R-02/084. October 2002.

Singh, A. (2004) Estimating the exposure point concentration term using ProUCL, Version 3.0. Presented at Annual Meeting of the Society for Risk Analysis, Palm Springs, CA, December 15.

Tannenbaum, L.V. (2003) Can ecological receptors really be at risk? *Human Ecology and Risk Assessment* 9:5–13.

Tannenbaum, L.V. (2005) A critical assessment of the ecological risk assessment process: A review of misapplied concepts. *Integrated Environmental Assessment and Management* 1:66–72.

Tannenbaum, L.V., Johnson, M.S., Bazar, M. (2003) Application of the hazard quotient in remedial decisions: A comparison of human and ecological risk assessments. *Human Ecology and Risk Assessment* 9:387–401.

US EPA (1989) Risk Assessment Guidance for Superfund. Volume I: Human Health Evaluation Manual (Part A), Interim Final. Washington DC: US EPA. EPA/540/1–89/002. US Environmental Protection Agency.

US EPA (1990) National Oil and Hazardous Substances Pollution Contingency Plan, Federal Register 55(46):8666–8865. US Environmental Protection Agency.

US EPA (1992) Intermittent Bulletin Volume 1, Number 1, Supplemental Guidance to RAGS: Calculating the Concentration Term. Office of Solid Waste and Emergency Response, Publication 9285.7–081. US Environmental Protection Agency.

US EPA (1993) Wildlife Exposure Factors Handbook, Volume I. EPA/600/R-93/187a. US Environmental Protection Agency.

US EPA (1997) Ecological Risk Assessment Guidance for Superfund: Process for Designing and Conducting Ecological Risk Assessments, Interim Final. EPA/540-R-97–006. US Environmental Protection Agency.

US EPA (2001) ECO Update, Intermittent Bulletin, The Role of Screening-level Risk Assessments and Refining Contaminants of Concern in Baseline Risk Assessments. EPA 540/F-01/014. Office of Solid Waste and Emergency Response, Publication 9345.0-05I. US Environmental Protection Agency.

US EPA (2002) Guidance for Comparing Background and Chemical Concentrations in Soil for CERCLA Sites. EPA 540-R-01-003. US Environmental Protection Agency.

US EPA (2003a) Ecological Soil Screening Level for Aluminum, Interim Final, Office of Solid Waste and Emergency Response Directive 9285.7-60, Washington DC. US Environmental Protection Agency.

US EPA (2003b) Ecological Soil Screening Level for Iron, Interim Final, Office of Solid Waste and Emergency Response Directive 9285.7-69, Washington DC. US Environmental Protection Agency.

US EPA (2007a) Advice to EPA on Advancing the Science and Application of Ecological Risk Assessment in Environmental Decision Making. http://yosemite.epa.gov/sab/sabproduct.nsf/7140DC0E56EB148A8525737900043063/$File/sab-08-002.pdf

US EPA (2007b) http://yosemite.epa.gov/sab/sabproduct.nsf/MeetingCal/BF1AC7EBFA2A8A998525723B-006D0CB4?OpenDocument.

US EPA (2008) Guidance for Developing Ecological Soil Screening Levels, Office of Solid Waste and Emergency Response, Directive 92857.7-55, Washington, DC, SSL values on line at: http://epa.gov/ecotox/ecossl/index.html. US Environmental Protection Agency.

US EPA (2011b) Ecological Structure-Activity Relationships. http://www.epa.gov/oppt/newchems/tools/21ecosar.htm. US Environmental Protection Agency.

Wenning, R.J. & Apitz, S.E. (2012) Ecosystem Services: Protecting the Commons. Integrated Environmental Assessment and Management. 8:395-396.

8 A new ecological assessment paradigm for historically contaminated sites: Direct health status assessment

The earlier chapters were intended to set the stage for the assembly of an alternative approach to ecological assessment for conventional hazardous waste sites. Specifically, the chapters hinted at what the elements of an alternative paradigm might be while they were also allocating time to reviewing the flaws of the present ERA process. Imperative to the current discussion is a proper understanding of the context in which a new paradigm is being proposed. We live in an age where new approaches to doing things come about not because we know them to necessarily be any better than what we presently have, but because people just want to see or try something different. With the proposed paradigm, this is most definitely *not* the case. The new paradigm comes about with a deliberate intent to improve our understanding of site ecology which may very well bring along with it the considerable fringe benefit of informing that contamination does not pose undue (or any) stress to site receptors. A trickle-down benefit of learning that receptors are not stressed is discovering that additional studies and cleanups are not needed.

What is being proposed is not a modified paradigm, but rather a completely new one. In fact, there are no elements of the present ERA process/paradigm that are common to what is being proposed. In the spirit of what has been presented until this point, an unwillingness to ponder the design of a completely different paradigm will all but ensure that what is described in this chapter and the following one will not be given due consideration.

The present-day and proposed paradigms radically differ from the very start, that is, with regard to the purpose of engaging in assessment work. Because so much time has elapsed at contaminated sites by the time we engage in study, asking about what toxicological effects could arise in site receptors is passé. The new paradigm therefore heralds a shifting of tenses, no longer asking "What

Alternative Ecological Risk Assessment: An Innovative Approach to Understanding Ecological Assessments for Contaminated Sites, First Edition. Lawrence V. Tannenbaum.

effects could (still) arise at this site?" but instead asking "Have effects occurred at this site?" It is imperative that this distinction be appreciated. The previous chapter briefly touched on the findings of the Coeur d'Alene River Basin ERA (CH2M Hill & URS Corp. 2001), and the findings are again useful here. That ERA found spotted sandpipers to have double and triple digit LOAEL-based HQs for several chemicals. Under such conditions, no spotted sandpiper could possibly survive, nor could any other bird species (despite the fact that diets and behaviors of other birds might substantially differ from those of the spotted sandpiper). Yet, the river basin was not described as being devoid of all bird life. We also know that spotted sandpipers populated the river basin at the time the assessment was conducted, for the assessment would not otherwise have considered them; we are not in the habit of assessing species that we only wish or imagine were present at a site. Further, prominently placed among several commonly agreed upon receptor selection criteria is that receptors have a relatively high likelihood of contacting chemicals via direct or indirect exposure (Suter et al. 2000). With spotted sandpipers in abundance, the reported HQs of the Coeur d'Alene assessment need to be taken to task, for they have surely provided a great misread on the ecological site dynamics. If sandpipers were evident after more than a century of exposure to extreme metals concentrations, it was wrong to investigate the prospects for continued species survivability altogether. If anything, the excessive HQs (despite their unwieldy magnitudes) indicate that sandpipers can well tolerate the high exposures to metals they encounter. If the concentrations of metals were as damaging as they were made out to be, sandpipers at the river basin would have died off long ago.

If spotted sandpipers and numerous other bird and mammal species were observable at the Coeur d'Alene River Basin with all of its excessive mine tailings, and after a century or more of exposure, "risk" assessment wasn't needed. Potentially though, the river basin wasn't supporting as many animals of each kind as it should have been, or as many as it had supported in previous times. This suggests that it could have been more appropriate to census species at the basin (than to have run a food-chain model) to establish if the contaminated environment was limiting to receptors in a more subtle way. In theory then, it is suggested that a vastly improved ecological assessment paradigm for contaminated sites might not require anything more than censusing a limited array of species, and reviewing the data in a context of site capability to support as many animals of a given species as it should. Potentially there could be two census comparisons, with the primary one pairing up site animal numbers (possibly as catch/unit effort) along with age distribution and sex ratio statistics, with those of a nearby habitat-matched property. The second comparison would pair up site animal density figures with those reported in the open literature. There are difficulties, however, with simply conducting animal censuses, and while censusing can be a functional component of a proposed alternative paradigm, the reader should not think that censusing can constitute all of it. For a proper census, a relatively nearby, highly habitat-matched non-contaminated area needs to be selected, and such may not exist. Even if the matching were respectable, opponents of an approach that intends to rely on comparative population counts (of site and reference location) as a means of assessing receptor health can be anticipated to question the reference location choice each time. There is a complication too for the prospect of reviewing site census figures with those found in the literature. Chances are that the literature will not furnish density figures for the specific locale of interest. How many density studies are there for opossum in central Alabama?

While censusing may not constitute an ideal or practical replacement scheme for ERA, it is not without recognized utilities. The beauty of simply setting out to count animals to determine if sites are supporting the biota they should, is that toxicological study as it proceeds today in support of ERA (with all of the uncertainty and non-applicability it entails) can be dispensed with. Given toxicology's track record in support of ERA, ecological assessment won't be any better off when it attempts to describe the degree to which receptors might be consuming greater quantities of contaminants than are thought to be safe. A passing consideration of censusing as a stand-alone assessment tool has a second benefit. For the common case, it illustrates how absolutely non-applicable wide-ranging receptors are in the context of contaminated sites. Thus, we should hope that contractors tasked with censusing fox or long-tailed weasel at a 15-acre site and at a comparable pristine area would come to the realization that such work was purposeless; for all of their efforts, they wouldn't be expected to spot even one animal of either type. The greatest benefit of a consideration of censusing is that complications involved with such work necessarily relate to the shaping of a replacement ecological assessment paradigm that *is* workable. Because ERA practitioners have become complacent with desktop calculations, they are disinclined to conduct the tasks that can yield the information that is needed. The complications associated with censusing begin with a general perception that this work is mere drudgery. Beyond this, that censusing may require multiple visits to the field, involve one-of-a-kind techniques, and necessitate the deployment of specialized or dedicated equipment, are complications not working in our favor. As with censusing, however, effective ecological assessment can only materialize with a willingness to overcome our disinclinations (and our laziness). In reviewing censusing as a lead-in to the framework of an alternative assessment paradigm, we need to recall that assembled censusing data is only utilitarian if it is known how many fewer species representatives, relative to some standard, there needs to be to signify impact. Similarly, a technically sound basis for the number and types of species to be sought out when applying any other useful task must be furnished. These matters are not easily decided, but for the design of a paradigm that can reliably indicate if troublesome toxicological effects of concern have taken hold, they must be squarely addressed.

A review of the drawbacks of present-day ERA (Table 8.1) can greatly assist in communicating an effective new ecological assessment paradigm. Although the intent of the depicted process is on assessing risk, this is never accomplished for the reasons discussed earlier. The table also alludes to the great propensity for the process finding sites that, to its thinking, have at least one ecological problem. Most importantly, the table illustrates that there is no exit strategy for the ERA process. Earlier, after mentioning that the present-day and proposed paradigms differ in their very purpose, it was said that there are no elements of the present ERA process that are common to what is being proposed. It is time to see that this is true and to review the elements of the proposed paradigm that is correctly titled "direct health status assessment."

The direct health status assessment paradigm is necessarily a top-down approach. Onsite biota that are observable and that are free to be culled from the wild, provide the luxury of a window into how successfully, or perhaps unsuccessfully, animals fare when living amid contamination. The objective is to assign a health ranking to animals that we can actually touch, and not to project what is going on inside them based on what our knowledge of pharmacology and toxicology might tell us. Direct health status assessment begins with a fundamental underlying premise. Given the decades that have elapsed, toxicological effects of concern have had every opportunity to arise in site receptors. If we do not detect effects upon very close examination, we should not suspect that latency for the expression of effects is still a possibility. In a figurative sense, our best foot forward would have us picking up the very animals occupying contaminated sites, bringing them indoors to a doctor's office of sorts, setting them down on an examination table, and subjecting them to the equivalent review a human patient receives – looking down the throat, listening to the heartbeat, tapping the knees with a rubber hammer while the legs dangle over the table edge – and only doing so, of course, where it is clearly known that certain recorded measurements are clear indications of disease/critical biological function compromise, or accelerated senescence/mortality. In keeping with this figurative example, there would be no purpose to taking an animal's chest x-ray unless it was known how much larger or smaller an internal organ would have to be to indicate ill health.

The direct health status assessment paradigm recognizes that a site receptor reflects the influences of the many and varied stressors to which it is subjected, a fair number of which are not chemical in nature, and have never been incorporated into ERA analyses (e.g., noise, land vibration, ambient temperature). The paradigm also recognizes that effects, should they be detectable, may only arise because of the combined influences of a site's independent stressors. There is more. ERA has elected to trivialize the chemical inputs of terrestrial receptors through the

Table 8.1 Problem-laden elements of the current ERA process.

Current ERA process component	Difficulties posed by the process component
Decide that a site might pose risk because it bears contamination	1 Flawed premise; with sites having been contaminated for decades, it's too late for estimating or calculating risk (i.e., the potential for health effects to arise)
	2 Sites are often too small to house a sufficiency of receptors to justify study
Screen detected contaminants to determine which should be carried through an assessment	1 Consideration not given to the effects of chemical mixtures
	2 Consideration not given to chemical form (i.e., chemicals not speciated; chemical aging effects not studied)
Select receptors to model (that hopefully represent others)	Selected species are often not spatially relevant to the site (i.e., not enough animals of any one kind are present and/or animals don't contact a site sufficiently to have a toxicological response elicited)
Calculate chemical-specific expressions of the degree to which a safe intake rate is/is not exceeded.	1 Only ingestion considered as an uptake route
	2 Assumes that site-ingested chemicals behave as they do in laboratory animal dosing studies
	3 The concept of chemically integrated diets (or exposures in the case of EEQs) is not dealt with
Conclude there is a potential for risk where HQs > 1	1 HQ/EEQs are not risk expressions
	2 HQs/EEQs in any format are only screens; they are too preliminary and imprecise to act upon
	3 HQ method limitations are overlooked*
Conduct additional studies: toxicity tests, bioaccumulation studies, seek out improved source and ingestion terms	1 Additional studies don't relate to the study question
	2 Recalculated HQs bear the same limitations as originally computed ones
Calculate cleanup numbers for HQs > 1	1 Unacceptable risk is not demonstrated with HQ > 1
	2 Overlooked: pristine site conditions can have HQ > 1
Monitor the site rather than implement remedial action (for fear of doing more damage to environment by remediating)	1 The practice throws into question why HQs were computed
	2 It's too late to be monitoring
	3 Monitoring criteria are arbitrarily set

*HQs: (1) are not measures of risk, (2) do not indicate the fraction of a population that can be affected, (3) are often unrealistically high and toxicologically impossible, (4) often exceed 1.0 for the most minute of concentrations (inorganics), (5) are not linearly scaled, (6) are not linked to a temporal component (i.e., a HQ of 10 means the same thing for a site that' s been contaminated for 5 years or 500 years).

inhalation and dermal contact uptake routes. Although these two routes may admittedly only be minor contributors to the total chemical loading doses of receptors in the wild, the impetus for discrediting or discounting them likely reflects ERA's inabilities to understand them and furnish toxicity values for them. Although there may be no need for partitioning estimated exposure levels or potential effects by individual uptake routes, the judgments that direct health status assessment are able to render (of the health of site receptor populations in the face of contamination) reflect the cumulative effects of all operative pathways.

Operationally, direct health status assessment is most different from ERA with regard to considerations of site chemistry. With the former there is no chemical screening at the outset, as there is no need to identify a list of potential problem posers in order to proceed. Instead, it is sufficient just to know that a site is genuinely contaminated, and as long as this is established in one or more reputable reports, the new paradigm can be applied. As a deliberate maneuver (and by no means through oversight) site chemicals are also *not* reviewed for their relative toxicity or for the organs or organ systems they are known to attack. Such information isn't particularly relevant given the new paradigm's awareness that site receptors do not experience individual chemical exposures one-at-a-time. Moreover, for the proposed paradigm such chemical review is premature, because the first order of the day for the paradigm is to establish if anything is "wrong" with site receptors – with "having something wrong" denoting an inability to have an essential biological function proceed normally.[1] It is a distinct possibility that for the overwhelming majority of sites, receptors do not have anything "wrong" with them. Where this is the case, it will become evident retrospectively that a chemical toxicity review was not necessary. Where compromised health *is* identified though, chances are that identifying the problem-posing chemical(s) will not stand as a great challenge; ERA practitioners will have likely formulated a hunch about these well before any formal assessment efforts got underway.

Figure 8.1 illustrates the proposed paradigm's alternative assessment orientation by depicting it alongside the conventional ERA process. As the figure first makes clear, with the new paradigm it is assumed that sites of concern are truly contaminated, and only the most minimal energies are expended on verifying same. With direct health status assessment seeking to uncover compromised key biological functions in site receptors (e.g., reproduction), it would be a complete waste of time to apply the proposed paradigm where such outcomes are not realistically anticipated (i.e., at clean sites, or perhaps at sites that have construction debris lying about). Importantly, and as mentioned earlier, since direct health status

[1] As was explained in Chapter 4, measureable differences in physical attributes of site animals (e.g., voles with reduced liver enzyme levels; salamanders with shortened tails) do not of themselves make for cases of "something wrong". For measureable differences to carry such weight, *a priori* it would have to be known how much of a measured difference signifies challenged health.

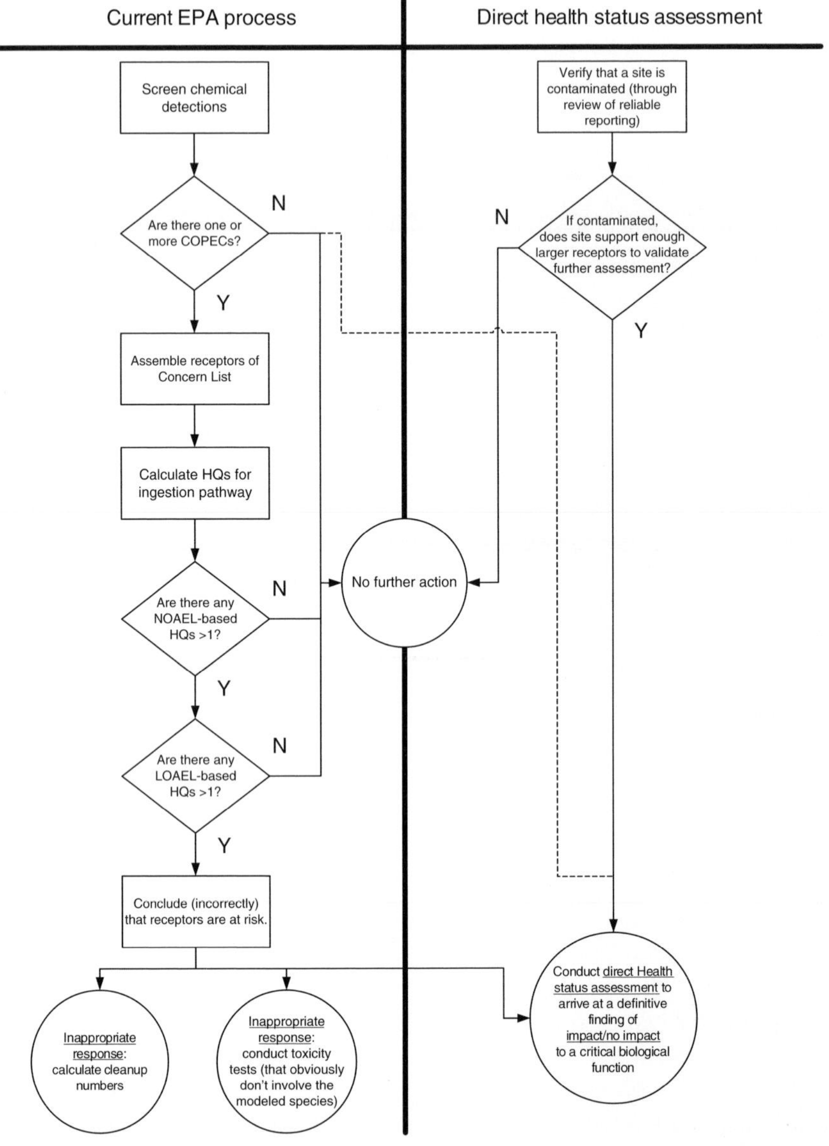

Fig. 8.1 The diagram points to the conventional process ultimately defaulting to direct health status assessment as a matter of need in almost any circumstance. The conventional process invariably concludes that one or more receptors are at risk, and chooses poorly at that point. The alternative approach (right side) knows to opt for direct health status assessment from the very start so that a definitive finding can be made.

assessment doesn't generate chemical-specific findings, there is no need to characterize a site's pattern of contamination very well (e.g., to know where maximum detections are located or where certain chemicals co-occur); again, it is enough to know that there is contamination present with which site biota are interacting.

Direct health status assessment is markedly simplistic in terms of flowchart steps. To the extent that we may speak of the conventional ERA and proposed paradigm methods running in parallel, in place of contaminant screening in ERA, is *receptor presence-based screening* (as first described in Chapter 3). Based on our knowledge of densities of those species for which site remedial efforts could realistically proceed (e.g., deer, lynx, owl), the user, pursuant to this screening procedure and flowchart step, should be primed to discover that a goodly number of sites will not be found to be in need of assessment. (In this context, the reader should recall Tables 3.1 and 3.2 so as to maintain a familiarity with defensible minimal site sizes for the consideration of ecological assessment work for mammals.) As per the flowchart, without the required "parent material" to allow an assessment to take hold, the proposed paradigm terminates abruptly with No Further Action. As explained further later, direct health status assessment necessarily involves the collection of specimens of the more expendable sentinel species (e.g., field mice). The user should understand that while sentinel species may be present in high numbers at a given site, the larger species that are the true intended objects of ecological assessment efforts may not be. To assist the reader with concretizing this point, we can consider the case of white-tail deer in a forested habitat with a relatively hefty abundance of 40 animals/km^2. Recalling that the majority of Superfund sites are 20 acres (8.1 ha) in size or less, for a contaminated parcel of this size within the census area, fewer than four deer would be expected to be present – quite obviously, not enough animals to worry about.[2]

Where receptor presence-based screening indicates that a site *does* support a sufficiency of those receptors we aim to protect, direct health status assessment proper is initiated. Generically, one or more key biological measurements of a sentinel species that co-occurs at the site in question and at an equivalent nearby, non-contaminated, habitat-matched location are compared. An indispensable element for the comparison is the knowledge of the degree to which a biological measurement needs to be shifted (i.e., changed from the condition at the non-contaminated location) to be indicative of compromised biological function. This of course also presumes that the direction of favorability/non-favorability of a measurement is known. Thus, if compromised respiration was of interest, it would first be established that *decreased* (and not increased) respiratory volume, attributable to chemical site exposure, was symptomatic of respiratory injury.

[2] Of course, the non-necessity for considering deer for assessment is further demonstrated by taking into account the species' average home range, which is generally in excess of 600 acres.

Critically, the essential measurements comparison of site and reference location animals does not merely reduce to the reporting of an absolute difference between population means, perhaps expressed as a percent. Rather, direct health status assessment is predicated on comparing the computed absolute measurement difference with an empirically derived threshold-for-effect. An exaggerated hypothetical example will illustrate this.

It is decided that the ability for site mammals to move efficiently is the dominant concern at a contaminated site that has been mothballed for 30 years. In arriving at that position it was suspected that two site contaminants had the potential to reduce nail length on the front feet of small mammals. Insightfully, too, it was recognized that while perhaps none of the other site contaminants induce a short nail effect when acting individually, they can do so when acting in concert with others. Shorter nails could mean that small mammals fail to gain sufficient traction with the site substrate, resulting in an inability to run as fast as they optimally could when exhibiting avoidance behavior. Consequently, small rodents with compromised toenail-length would be at an increased risk of predation. A depletion of small rodent numbers through site-mediated excessive rodent predation would indicate that there is a malfunction in the larger terrestrial site ecology. To facilitate the field-based analysis, 20 specimens of each of two small rodent species that occur at both the contaminated site and a nearby habitat-matched property, were trapped in the field and had their front nail lengths recorded (before being released to the field). When population means were compared, both species at the contaminated site were discovered to have nails that were, with statistical significance, 18% shorter than those of their reference location counterparts. Importantly, the established reduced nail length threshold-for-effect, of 25%, was not exceeded.

Although the example bears the makings of utilitarian direct health status assessment, three critical aspects of the study arrangement require immediate attention. While theorizing can be appreciated, the decision to collect the nail length data should only have proceeded if it was definitively known that shorter nail length is a barometer of substrate traction and running speed. Assuming such is the case, for any good to derive from the field study, and prior to relying on the nail length measure, it would have to be determined through empirical research how much shorter nails need to be before a statistical difference in running speed is evident. Finally, the degree of running speed reduction that results in rodents being predated upon more than is ordinarily expected would have had to be established. The example then is intended to underscore the types of prerequisite, toxicologically based information that must be assembled before direct health status assessment can be applied in earnest. The intention was also to underscore that perhaps with only one exception (i.e., the methodology reviewed in the next chapter), the data to support direct health status assessment does not presently exist.

The intent of direct health status assessment is to arrive at definitive determinations, or as definitive determinations as are possible (given certain insurmountable constraints with which we must live) about population health for species that genuinely concern us. The example was designed to galvanize this point in a peculiar way. Although certain elements of the example are relevant, others are most definitely *not* workable. Thus, the proposed paradigm is not intended to link effects in sentinel species (to include the erasure of them) to spurious and suspected-only trickle-down effects in selected receptors of concern. The paradigm therefore, cannot effectively allow for relating reduced numbers of an intended sentinel species (as in the projected reduced rodent numbers achieved through stepped-up predation) to larger-scale, ill-defined ecological effects (as in the suspected "site ecology malfunction" mentioned before). Just as the sentinel species should be defensible as a surrogate for one or more species that genuinely concern us, so too, the essential somatic measurements to be collected for driving the assessment scheme must be reasonable parallels for the species that genuinely concern us. In the example, if we are unsure about the functional relatedness of front feet nail length in mice and claw length in fox, or perhaps pronghorn hoof design, the comparative assessment is not supported.

We should return to the example, albeit now somewhat modified, to draw out several essential points. Assuming (a) reduced nail length *is* a barometer of reduced avoidance response, (b) toenails in mice are highly functionally aligned with equivalent tissue of the receptor of concern (say, cottontail), and (c) nails that are shorter by 25% from the control rate constitutes an established threshold-for-effect, we can conclude through extrapolation that cottontail at the hypothetical site are not harmed. As to the mechanics of the extrapolation, field mice likely have a greater degree of soil contact and soil chemical exposure than do cottontails given the former's relatively smaller home range, more frequent burrowing activity, and smaller size. If the threshold-for-effect in mice wasn't exceeded, by extension it was not exceeded as well for the cottontail.

The workings of the paradigm, even as demonstrated with this crude example, should be appreciated, and particularly in relation to the present ERA process.

- Although there is uncertainty associated with field mouse-to-cottontail extrapolation, the gain to the science in having collected defensible and utilitarian biological information from actual site receptors cannot be underplayed. To secure due appreciation for the new paradigm's capabilities, the reader might wish to consider the only other alternative we know of, and that could have been applied in this (hypothetical) site instance. That is, soil contaminants would have first been reviewed, with several determined to be retained for assessment. Ingestion-based HQs, based on mouse- or rat-derived TRVs, would have been developed for the cottontail, and at least some of these would have undoubtedly exceeded the unity threshold by some considerable degree. It would have been concluded that cottontail are at risk, and cleanup numbers for the protection of cottontail would have been derived on a chemical-specific basis through back-calculation to HQs of 1. Relative to the direct health status assessment design, the standard ERA approach pales. No site receptors were ever observed, much less handled or examined with ERA, and only one contaminant uptake route was considered. At a time when effects (here, reduced nail length) should have long-time arisen, the ERA process talked to the future chances of nail shortening. This latter point highlights the most profound difference between the two approaches. Direct health status assessment is designed to arrive at a definitive finding and it is only applied where the scientific supports are in place to ensure that this will be the case. The ERA process (as depicted in Fig. 8.1) is indeterminate, subject to groping at additional means to project the chances for toxicological effect expression, and unable to arrive at assessment closure.

- Direct health status assessment centered about the field-mouse as a sentinel species should not be subject to criticism because of the extrapolation involved. [Ideally of course, cottontails as the assessment endpoint species (and not mice) would have been culled from the affected site and its corresponding reference location. Logistical concerns such as the time-consuming and labor-intensive nature of the animal collection work though, likely fed the decision to instead trap mice as cottontail surrogates.] While extrapolation necessarily introduces uncertainty to any assessment scheme, the extrapolation that was employed in the hypothetical instance is undeniably a truer one than that which typifies routinely conducted ERAs, because actual site animals were studied. Here the reader must appreciate that regulators and other stakeholders have no qualms about declaring that receptors are at risk and that sites consequently require remediation, when the *standard of proof* (if we may) is limited to interspecies extrapolation for situations involving no animal collection. If we can be so bold as to "convict" sites when relying on a far less imprecise algorithm (i.e., one that in addition to not involving site animals, reviews the estimated chemical doses of artificially exposed laboratory species), we should certainly be as willing to decide the health status of directly assessed species, although this approach too involves extrapolation.
- In a context of the proposed paradigm's fundamental underlying premise, the ability to secure definitive determinations with direct health status assessment, and particularly where such determinations may be positive (i.e., that site receptors are not health-compromised), cannot be understated. In the example, the site's field mouse population had nail length on the front feet reduced by 18% relative to that of reference location field mice. It is only where nail length is reduced by 25% though, that a case can be made that locomotion is compromised. Hence it was concluded that just as mice were not experiencing

excessive predation, neither were site cottontails, the selected receptor of concern. Critically, there should be no allowance for users expressing nagging feelings about the assessment outcome, as in (a) the measured 18% difference in the direction of non-favorability being fairly close to the threshold, or (b) suspecting that in the out years, the 25% threshold would ultimately be approached and exceeded. Harboring such feelings at historically contaminated sites reveals an unwillingness to accept the fundamental underlying premise of direct health status assessment, simplified here to: decades-old contaminated sites are as bad as they'll ever be (in terms of having the potential to elicit serious toxicological consequences). Suggesting that decades-old contaminated sites that failed to demonstrate an effect need to submit to repeated demonstrations in still later years, is unfairly distrustful. It should go without saying that in those instances where conventional ERA and direct health status assessment were both applied, deference should be given to the latter.

It is easy enough to place direct health status assessment in its proper context. Ecological receptors at contaminated sites could be experiencing toxicological effects severe enough to threaten their survival. Our tasking therefore should be to seek out ways of detecting if somatic effects indicative of such extreme toxicological insult have come about. In so doing, we should not become side-tracked with observations of other somatic changes in site receptors that although real and statistically significant, are not indicative of the severe health challenges that interest us. Detecting the severe effects can be facilitated through either of two comparisons, one where empirical research establishes an absolute threshold, and the other where empirical research establishes a relative one. With the latter, after recording the same biological measure in site receptors and conspecifics of a nearby habitat-matched non-contaminated area, we employ knowledge of the percentage increase or decrease from the norm (i.e., the non-contaminated site condition) that signifies disease and death as anticipated exposure outcomes. The hypothetical small rodent nail length example would describe the comparative assessment case. Of the two comparison types, potential advantages attain to absolute threshold comparison-based determinations, this because receptors need only be collected from the contaminated sites. In addition to lesser levels-of-effort needed for securing field specimens, absolute comparisons are less destructive, recalling that more often than not an animal's life is taken en route to collecting necessary biological measures.

The reader should appreciate that the absolute and comparative biologically significant thresholds-for-effect that fuel direct health status assessment barely exist at the present time. This, in and of itself, is not a limitation of the proposed paradigm. As was briefly alluded to in Chapter 4, reasons for their scant existence (Tannenbaum et al. 2007; Paulose et al. 2012) include the demonstrated preference for and complacency with employing food-chain models as *de facto* health assessment tools, and practitioners not having considered or imagined alternative assessment schemes. Admittedly, procuring thresholds-for-effect to support the proposed paradigm necessitates labor-intensive work of a new variety. For our needs, it is most important to recognize that the standing body of dose–response literature is of very limited use in a direct health status assessment vein because of its repeated focus on observed statistical differences between test groups. Since direct health status assessment keys into biologically significant differences (and not statistically significant ones), the experimental design of studies to uncover these is necessarily different and considerably more elaborate. We should not be discouraged at the required initial outlay of time and other resources to assemble the underpinnings of the proposed paradigm. For now, the engaged and open-minded reader should consider that direct health status assessment might inform us that site receptors are not health compromised (as was previously mentioned). If this was the case, the time investment and research costs leading to the establishment of truly utilitarian biologically based thresholds-for-effect would pay for themselves many times over, because (unnecessary) site remediation costs wouldn't have to be realized. As the next paragraphs describe the etiology of a utilitarian threshold-for-effect (again with the aid of a hypothetical example), a practical consideration is the intent to have an engineered threshold that is as broadly applicable as possible. This can be achieved by stepping back from the site-specific level of organization to perceive all contaminated sites, regardless of the chemicals they present and the concentrations in which they occur, as bearing the capacity to impact the biota that utilize them. A de-emphasis therefore on known or suspected linkages of detected chemicals and toxicological effects, will ensure that the potential for certain toxicological endpoints to have developed will not be overlooked. Realistically we might not know of all the chemicals that, acting singly or in combination, trigger a particular effect. By way of example then, if altered behavior is a shared concern for regulators or for the project team as whole, this endpoint should be explored sooner than relying on a team's claim of fluency with the exhaustive list of chemicals that elicit altered behavior. The reader should consider here, the routine application of toxicity tests in ERA. While the tests are directed to highly specific endpoints (e.g., earthworm cocoon production over 28 days), application of the tests is never preceded by a checklist review to establish that at least one chemical present in the affected medium (e.g., soil) elicits a test endpoint. In a similar way for the considered and proposed paradigm, the toxicological endpoint that concerns us most is selected, and the effort is made to ascertain if the endpoint has been realized in site-exposed receptors.

The intent to design a working threshold-for-effect begins with a modification of what would be a problem formulation statement found within a conventional ERA. In the hopes of facilitating the development of a versatile status assessment tool, a statement such as "Hawks at Operable Unit 3 of the Jones Trucking Site are potentially at risk for reproductive effects from their consumption of onsite lead-contaminated food" would be adjusted to "Are onsite birds experiencing reproductive effects?" With the question formulated, thought is then given to selecting a data-type, collectible at any site that supports birds, that when shifted in a non-favorable direction (at the contaminated site) should be indicative of reproductive compromise. For illustrative purposes only, we suppose that thinner eggshells might be developing in birds at many a contaminated site, this because of the interplay of several contaminants that are routinely detected in analytical soil work. The objective becomes one of establishing how much thinner an eggshell needs to be (from the norm) before failure ensues (i.e., when development is arrested because the shell cracks under the weight of nesting parents tending to their brood). After learning of that critical difference, the goal would be to collect eggs from contaminated sites and their paired non-contaminated reference locations, for a structured comparison of eggshell thickness among and between the discrete populations. The data review would establish whether or not eggshells have become so thin as to indicate lesser reproduction (occurring through the production of fewer successful clutches or smaller-sized successful ones).

For the hypothetical case, the empirical research to lead to the desired threshold would begin with cautiously selecting a test species that can serve as a surrogate for two other birds (i.e., the species in the field whose eggs would be removed for analysis, and the receptor of concern). (For obvious reasons, the eggs of hawks and other larger, top predators would never be harvested.) Ideally, the test species would have a similar behavior and/or feeding design (e.g., granivorous, carnivorous) to the two other birds. At a minimum, the reproductive strategies of the surrogate and the receptor of concern species, in terms of clutch size and number of clutches per year, would be similar. Experimentally, discrete groups of the test species in the laboratory would be administered variable doses of a controlled substance such that a range of eggshell thicknesses would be produced. While artificially incubating the eggs, gentle pressure simulating the nesting parent weight would be applied to the top of the eggs, and the shell thickness that cannot sustain the applied weight during incubation (i.e., the absolute threshold) would be ascertained. A relative threshold-for-effect would be crafted by using the identified absolute benchmark to express the percent thickness reduction from the norm (the study's control population) that triggers shell failure.

With the experimentally derived thresholds now secured, a remaining task for direct health status assessment implementation is the selection of the field surrogate species. Certain qualifications are obvious: there should be a sufficiency of nests onsite and offsite to allow for a reasonable comparison in terms of sample sizes, and the species should have both a relatively high degree of spatio-temporal site contact and direct soil contact. In addition to maximizing opportunities for the effect of interest (thinned eggshells) to have been elicited, the high contact rates allow for effective extrapolation as in the earlier mouse-to-cottontail illustration; if the eggshells of birds that spend most of their time at the site don't contravene the minimum shell thickness threshold, the same is true for the receptor-of-concern species that does not avail itself to having its eggs measured. Although not an essential requirement, the ability to collect data other than that needed to support comparisons involving biologically significant threshold use, is desirable. The latter can serve in a corroborative capacity, but the intent of direct health status assessment is not to have less speculative and less proven data displace developed and defensible thresholds-for-effect. In the hypothetical example, nest egg counts for the surrogate species should by all means be recorded while collecting the eggs that will submit to shell thickness analysis. In that case, however, where site eggshells are not found to have been thinned to the point of exceeding a reproductive standard, but surrogate species egg counts were statistically lower onsite, the determination of satisfactory reproduction for the receptor of concern would necessarily stand.

Of course, the hypothetical example is not workable. Perhaps the example was supplied because unusual or forced cases can often make for effective teaching tools. Non-workability in the example probably first begins with the unlikelihood of a contaminated site furnishing enough nests to support the sampling of the proposed assessment scheme. Presumably a single surrogate species from the field is sought so as to circumvent the complication of variable manifested eggshell thinning with species. A precursor to the application of the proposed paradigm then, is the review of nest densities for intended field surrogates. Such an exercise would recall the sizes of hazardous waste sites that seemingly necessitate our attention. This would trigger considerations of how many receptors of concern make an assessment worthwhile. In the example, assuming it would be fair to compare eggshell thickness of Henslow's sparrows and hawks (itself a complication of sorts), the number of hawks potentially at stake should be ascertained. This, of course, takes us back to Fig. 8.1's first (and only) flowchart decision point: have we enough site receptors of concern (i.e., representatives of a given species) to sanction health assessment through the use of field surrogates or otherwise? The flowchart decision would bring forward considerations of the home ranges of hawks, which would in turn, recall the commonly encountered difficulty of wide-ranging receptors insufficiently contacting sites to the point where unhealthful effects could reasonably be anticipated to set in. The example is probably unworkable too because of the labor-intensive nature of the required data collection. Aside from the complications associated with locating and accessing nests, with the sampling window (for observing unhatched eggs) being so particularly narrow, the collection work would be rather undesirable. If the bird assessment scheme was changed to a simpler nest egg count comparison, assuming of course, that it had been earlier ascertained how much smaller clutches need be to compromise reproduction (by no means a straightforward determination), the situation would be no different.

The engaged reader should not become discouraged with the complexities associated with developing thresholds-for-effect to support the new paradigm. In a sense the tasking is simplified because of the example's handiness, namely its alluding to the reality of there probably being only a limited number of potentially valid and utilitarian threshold development possibilities. Given the work involved in bringing a threshold online, the greatest concentration at the outset should be in the choice of toxicological endpoint to be reviewed. Alternatively stated, it would be best to leave alone the toxicological changes that some suspect bear on the "quality of life" (i.e., what we might think is bothersome or unhealthful for an animal). While ERA practitioners might obsess with certain biological measures that distinctly differ between sites and reference locations, it is pointless to pursue them if we remain unclear as to any health impacts they may intimate. The reader should appreciate that developing thresholds may be more difficult for some animal groups than for others, and also that it may be possible to intelligently extend the findings of the applied threshold(s) of one group or phylogenetic class to another.[3] Thus for terrestrial sites, where only mammals and birds concern us,[4] it would be worthwhile to develop and apply direct health status assessment for representatives of just one these two classes if that is the best that can be done. Developing biologically significant thresholds-for-effect for aquatic sites presents its own share of challenges. Potentially the challenges are of a greater magnitude than for terrestrial sites, and perhaps this book has provided fewer aquatic site examples because of this. Nevertheless, to bring ecological assessment for aquatic sites to a higher order, our present efforts must be seen for what they are. To date, although recommendations to implement sediment remediation have been commonplace, these have not been accompanied by demonstrations of aquatic site ecology having been compromised[5] to the point of needing this type of intervention. With decades having elapsed though, and with certain aquatic species continuously and hopelessly surrounded by a contaminated envelope of sorts, there is every possibility that aquatic health has come to be corrupted.[6] As with terrestrial assessment, en route to developing an aquatic threshold-for-effect, a valid endpoint of concern needs to be selected. Thus, we should be

[3] To be demonstrated in the next chapter.
[4] Although efforts have been made to craft TRVs for amphibians and reptiles, HQs resulting from their use are even less defensible than those computed for birds and mammals. Additionally, other than the direct health status assessment approach discussed in this chapter, for the ERA arena, there are no plans to assess any animal group, terrestrial or aquatic, with a mechanism other than a HQ screen.
[5] A changed and a compromised ecology are two vastly distinct phenomena.
[6] The previous chapters have argued to the contrary; that the extensive exposure duration should have more than guaranteed that observable health effects are present.

able to articulate what offset element/aspect of an aquatic system is troubling to us because we know it to be indicative of an impact requiring attention (sooner than later). Doing so is more challenging than we might realize. We are fortunate though, that we are able to identify some choices that are clearly unworkable because they insufficiently inform of a health or survival threat. These would include comparisons of sediment contaminant concentrations and tabular sediment quality protection benchmarks (where exceedances of the latter would be revealed), and observations of a waterbody's bottom being predominantly or exclusively populated with pollution tolerant macroinvertebrates.[7]

With the essentials of the proposed assessment scheme having been reviewed, the reader's attention is turned to the matter of prospects for the scheme gaining acceptance within the ERA community. Several challenges to the scheme are easily identified, and these begin with the anticipated uneasiness associated with adopting a decidedly different approach. Indeed the proposed scheme is directed to answering a different question than is ordinarily entertained. To sidestep the hurdle, it is suggested that every effort be made to have the conventional and proposed assessment schemes always run in tandem. Doing so will present opportunities for the leery to consider what the proposed scheme's intention actually is, and to foster appreciation for the scheme's site-specific elements. It is anticipated that some will be discouraged over the laborious nature of threshold-for-effect development. There are several ways to manage this challenge. Rather than have stakeholders balk at applying the proposed scheme, it should be explained that despite the demands of threshold development, the pay-offs of the proposed assessment scheme truly outweigh the efforts invested in assembling it. One of the proposed scheme's pay-offs to showcase, perhaps its most influential one, is the unique capability to characterize ecological receptor health in a context of chemical mixtures. Harping on the complexity and slow pace of development for thresholds-for-effect can also be effectively addressed by explaining, as was mentioned before, that in all likelihood only a handful of thresholds are likely to ever materialize. It could further be explained that a robust threshold-for-effect, one that might already exist, can subsume the concerns for multiple elements of a site's ecology. Thus threshold development in earnest may largely not have to be realized at all. Undoubtedly, the greatest challenge concerns the outcomes of scheme application for direct health status assessment. Wisely, the engaged reader should be prepared to find in the common case – if not in every case – that ecological receptors at contaminated sites are not health-compromised, this bearing on the fundamental underlying premise of the

[7] These choices *would be* workable if it was known that the findings or observations are barometers of ecosystem impact. If, for example, one or more exceeded sediment benchmarks was a reliable signifier of fish populations being notably reduced, the benchmark screening would be appropriate and utilitarian.

field-based status assessment scheme. (If site receptors were shown to be health-compromised, we would be held accountable to provide a working explanation for why the receptors were still around, and such might not be easily assembled.) Realistically for at least some sector of the ERA camp, "good news" reporting can be anticipated to not resonate well; with their biases evident, these individuals would prefer to find that ecological impacts are evident at contaminated sites.

The immediately preceding discussion gives rise to a conundrum, and still one more hurdle for the proposed paradigm to outstrip in order to gain acceptance. With all the conservatism that can be built into a direct health status assessment scheme that is best designed, an outcome indicating that site receptors are health compromised may never arise. When doubting ERA practitioners observe this, we would expect them to come forward to challenge the merits of the applied threshold-for-effect and the assessment scheme overall. Principally they would allege that the scheme is insufficiently sensitive to detect health compromise, but this allegation is insensitive to the reality that receptors might not be experiencing toxicologically imposed health effects. What if animals in the wild are more resilient to chemical stressors than we realize? To the extent that animals are truly so tolerant of chemical stressors, it would not be fair to require a demonstrated case or two of a threshold-for-effect under development being exceeded in a field application before acknowledging the threshold's value. Whose fault is it that nerve impulse conduction in a rodent isn't sufficiently slowed to make for a difficulty, that bird song isn't altered to the point that mating does not proceed, or that a fish population's intrinsic rate of growth isn't reduced enough to imperil it?

A parting consideration for the proposed paradigm brings us back to Fig. 8.1. As ERA practitioners surely know, there is a great propensity for conventional assessments to terminate with HQs > 1, a veritable state of limbo. This can be alleviated by moving onto direct health status assessment. In anticipation of the need to arrive at definitives,[8] it could make more sense to implement direct health status assessment from the very start; hence the depiction of same in the figure.

References

CH2M Hill & URS Corp. (2001) Final Ecological Risk Assessment: Coeur d'Alene Basin Remedial Investigation/Feasibility Study. URS DCN: 4162500.06200.05.a2.CH2M Hill DCN: WKP0041. Prepared for U.S. EPA, Region 10, Seattle, WA, by CH2M Hill. Bellevue, WA, and URS Corp., White Shield, Inc., Seattle, WA. May 18, 2001.

Paulose, T., Tannenbaum, L.V., Borgeest, C., & Flaws, J. (2012) Methoxychlor-induced ovarian follicle toxicity in mice: Dose and exposure duration-dependent effects. *Birth Defects Research (Part B)* 95:219–224.

[8] In terms of clear indications of site animals themselves presenting with compromised key biological functions, and where all contaminant uptake routes and all site chemicals are collectively considered.

Suter, G.W. II,, Efroymson, R.A., Sample, B.E., & Jones, D.S. (2000) Ecological Risk Assessment for Contaminated Sites. Lewis Publishers, Boca Raton, FL, USA.

Tannenbaum, L.V., Thran, B., & Williams, K.J. (2007) Demonstrating ecological receptor health at contaminated sites with wild rodent sperm parameters. *Archives of Environmental Contamination and Toxicology* 53:459–465.

9 Is RSA the answer to ERA?

Although not expressly stated in the earlier chapters, the intent has been to insinuate that many an ERA practitioner, and by all means to include regulators in this group, espouse a guilty-until-proven-innocent philosophy with regard to contaminated sites. Sites to proceed through RI/FS type proceedings are assumed from the first to be assaulting the ecology left and right, and sites might only be viewed otherwise, sometime after ERA practitioners have taken them to task. With this approach, simple metrics for contaminated sites that don't align with background or some other reference point are perceived to be grave tidings. Thus a finding that a metal exceeds a protection level for benthic macroinvertebrate species in seven out of nine sediment samples of a lotic ecosystem, can be expected to evoke apprehension of the-sky-is-falling variety. More often than not, the discovered misalignments will trigger resampling efforts, additional testing, and strategizing for remediation. Isn't it curious that HHRA doesn't initially assume the worst with regard to the potential for contamination to be posing the most unhealthful of chemical exposures to the individuals who live or work at sites going through their RI/FS proceedings? With HHRA, we are instead open to see what the computed levels of risk and hazard are. Going into an HHRA, stakeholders may be bothered to know that a medium (say, groundwater) is contaminated, and they may have their guesses as to the risk assessment outcomes, but stakeholders are content to live by a risk assessment's findings. If groundwater ingestion has an acceptable level of risk or hazard, it must be that chemical exposure point concentrations are not as high as were suspected, or that relevant site receptors do not ingest as much groundwater as was thought.

Why we suspect the worst for the ecology of contaminated sites and why we take ERA to task so much more than we do HHRA, are questions that are not easily answered. A matter of more immediate concern is that of ERA practitioners 'crossing the line', i.e., allowing themselves to be dismayed or feel affronted with learning that a site is not as harmful to the ecology as they thought it to be. We will not revisit the role played by bias in ERA doings,[1] but we *will* recognize the

[1] As in necessarily promoting the case that ecological sites are problematic so as to ensure job security.

Alternative Ecological Risk Assessment: An Innovative Approach to Understanding Ecological Assessments for Contaminated Sites, First Edition. Lawrence V. Tannenbaum.

need to alternatively approach ecological assessment for biological forms living in association with contaminated media.

In a nutshell, an alternative approach to ecological assessment asks only that a contaminated site be given its fair chance to defend itself – to be able to show that it does not jeopardize the quality of life, or worse yet, threaten the very survival of the species it houses. It is all well and good to want to see that measurements, test outcomes, and indices for contaminated sites parallel those of correspondingly clean areas, but each of these information types is far less than exacting for our needs. For ecological assessment, receptors can be thought of as being present at contaminated sites for the express purpose of informing us, if we study them correctly, of their health state, and they stand to inform us far more reliably than can any index or ratio. As with HHRA, ERA practitioners should be willing to accept the findings, whatever they may be, of an applied technically sound assessment scheme (effectively, direct health status assessment), and they should shun any decided expectations or predictions they harbor. For the purposes only of shifting the balance away from guilty-until-proven-innocent thinking, the reader would do well to consider that just maybe, the ecology of a given contaminated site might be fine.[2]

The previous chapter set forth direct health status assessment as an adaptable approach. It was left in the hands of the ERA practitioner to select an aspect of a key biological function worthy of assessment, to devise a testing sequence that would furnish a barometer for the function's efficiency, and to decide on the field surrogate species and the specific measurements to record in these. The engaged assessor intent on developing the health status assessment tools that can be so serviceable, has a wide-open area to explore. Not meaning to discourage the assessor from venturing out to assume these tasks, but it may be that direct health status assessment has already arrived at its zenith, at least as far as terrestrial sites are concerned. The reference here is to the Rodent Sperm Analysis (RSA) method (Fig. 9.1), developed by the Army, applied in support of various site assessment initiatives, prominently factoring into remedial decision making, and recently patented (US APHC 2009).

RSA's humble beginnings trace back to a site-specific Army ERA and a project team of about 20 who shared a willingness to try something new after having been convinced that the status quo approach to ERA at their site was a fruitless endeavor (Tannenbaum et al. 2003a). Specifically, a Phase II RI had resulted in HQs in the multiple hundreds for a suite of terrestrial species and a suite of soil-bound metals and explosives with which they were interacting (SAIC 1999).

[2] To be fair, ERAs and even the direct health status assessment approach do not assess "site ecology". At best, we assess various site receptors, and through our review of these, we indirectly speak to a site's greater ecology.

Fig. 9.1 Rodent sperm analysis in a nutshell: established (reproductive) thresholds-for-effect are used to assess the health of the very site-exposed specimens as they occur in nature. Tannenbaum 2010. (Reproduced with permission of American Chemical Society.)

A passionate appeal for the project team to break with tradition registered in a very positive way when the likely sequence of events to ensue was mapped out. Thus it was explained that the team could first look forward to the RI contractor reworking large sections of the report to address comments that numbered in the hundreds (although admittedly there was considerable duplicity in the collective comments, as different stakeholder agencies had spoken to similar points). Also in some 6 months' time, the team would be reviewing an RI version that presented substantially smaller HQs than those of the initial effort; through relaxing the conservatism built into the initial values, HQs of 400 or 500 would likely then fall into the 20–30 range. Such HQ magnitudes, the team was reminded, would remain fully unacceptable to the regulators who sat on the team. The HQs would be perceived to be unacceptable although they would (a) not be articulating risk, (b) be referring to individual chemicals, and (c) not be coinciding with a considerable library of recent information documenting plentiful numbers of multiple species within multiple phyla throughout the entire installation. The passionate appeal really hit home when it was further explained that with crafting lesser magnitude HQs in the next go-round, the contractor would effectively be using up all the known tools that could paint a rosier picture for the site. The team was reminded that 6 months down the road, only introducing narrative arguments[3] potentially held some promise for finding the site to not be problematic, and also, that narrative arguments never really settle the score in ERA negotiations. Ultimately what moved the team to break with tradition was their realization that the ERA effort was nothing more than a game, where the objective was to manipulate HQs such that they would end up at or very near to 1. The ERA effort was seen to be divorced from the real-world condition in that values in tables, as opposed to animals, were being looked at. It certainly didn't hurt matters any to have the installation's naturalist sitting on the project team. With his institutional knowledge of the site ecology (knowledge, it should be noted, that pointed to the site ecology being particularly robust), which mirrored that of his father who had served as his predecessor, he enthusiastically took to the prospect of field-truthing the implications of the computed HQs and encouraged the rest of the team to embrace the task.

Curiously at the time the field-truthing approach was suggested, the concept hadn't been well thought through, probably because it wasn't expected to be met with acceptance. How though, could it be demonstrated that site terrestrial receptors weren't health compromised, as several team members suspected was the case? As discussed in the previous chapter, animal censusing beyond just a very few species, would likely prove to be too demanding. Even more curious on the fateful day the project team agreed to implement field-truthing, was that the concept of a comparative assessment scheme for the sperm parameters of rodents of different sites, was all but theoretical and imaginary. The concept nevertheless, grew quickly into the formalized and patented method it is today, one that intriguingly has barely necessitated tweaking since it was initially assembled.

The convergence of several essential points, a number of which have surfaced over a good many of the earlier chapters, formed the framework of RSA. The first was the realization that we are severely limited with regard to the species that we are at liberty to cull from the environment. In the case of mammals, our only choice is the small rodent, and we must accept this. It is rather unthinkable that those who would set out to conduct field-truthing, where such work necessitates expending animals, could selectively remove species that are larger and less numerous than mice, rats, and the like (e.g., fox, beaver, mink). Animal care and use committees would certainly frown on the larger species collections. In the unlikely case where collection permissions were nevertheless granted, it is clear that animals could not be removed from their habitat for any length of time. If trapping was allowed in a rare case, only minimally invasive sampling (e.g., removing a small bit of fur, or drawing off a small blood sample) might be granted. As we have seen however,

[3] By way of example, to have text added to the ERA indicating that soil chemicals hadn't been speciated (e.g., for form [salt] or valence state) and noting that potentially, site chemicals were not as toxic as the failing HQs would make them out to be.

identifying chemical detections in tissue is not serviceable to us, for we have no means to relate such detections to health effects. Additionally, we have an alternative understanding of chemical exposure/bioavailability (as per Chapter 3); our interest lies not in knowing of chemical concentrations that may manifest in tissues, but rather in knowing that the animals we study "hail from" (i.e., run or swim through) the contaminated locations that concern us. There are other issues that would work against us were we to have the liberty to cull larger species. In the overwhelming majority of cases, their home ranges would far exceed the site size, this at a time when we would prefer to consider worst case scenarios, i.e., where the receptor has minimal opportunities to live beyond the site. The related animal density complication follows right on the heels of the home range one. The prospects for culling a sufficiency of specimens of a larger mammal species (perhaps 15–20) to support a meaningful comparative analysis, would not present itself; it would be inordinately time-consuming and labor-intensive to procure the necessary animal numbers. Taken from a different vantage point, the impracticability of working with certain larger mammals, would spell disaster once more. Considering all that is entailed in developing a utilitarian direct health status assessment scheme, it would not be prudent to develop one that is focused about animals that are found only in certain habitats, or that have less than a ubiquitous distribution across the US such as the fox, beaver, and mink mentioned earlier. To put this into context – not that we know of any attempts elsewhere to piece together direct health status assessment schemes – it would not be prudent to devise a direct health status assessment scheme centered about American badger (*Taxidea taxus*) collections. This species is solitary for most of the year, largely nocturnal, and occurs only through some two-thirds of the US, primarily in grassland and other treeless habitats.

The concept of mammal-based direct health status assessment being rodent-centered necessitates a somewhat expanded discussion.

- There will probably never be a time when mammals other than small rodents routinely, if ever, constitute the heart of direct health status assessment. That said, the ERA practitioner must come to the understanding that a rodent-based scheme can be wholly sufficient to support a rather definitive determination for all other mammals at a site (Tannenbaum et al. 2007). In the aftermath of an RSA application therefore, it would be inappropriate to comment to the effect that the (rodent-based) assessment work conducted must be termed incomplete and inconclusive, because the actual mammals selected as receptors of concern had not been evaluated. It is imperative to understand that such assessment work will never occur. Further, since with nary an exception, the present ERA process bases *its* remedial decisions for larger, high-interest site mammals on rodent work, it would be unfair to express misgivings about relying on RSA rodent measures now, and particularly when the rodents involved come from the actual sites under review. Importantly, an unwillingness to accept structured field-based rodent data (as of the RSA variety) is to communicate an unwillingness to see sites get closed out, and to admit that a better assessment scheme has probably come along.

- While the specific rodent species to submit to RSA for any given application may be ever variable in terms of phenotypic features (e.g., body weight, tail length, coat color, gestation period), the user should be trusting of the assessment scheme's efficacy nonetheless. Down to the specific elements of the relevant biology to be discussed (reproduction), there is no reason to suspect that thresholds-for-effect are highly variable among rodent species to a point where they are considered unreliable. Here it is important to appreciate the previous chapter's hypothetical consideration of a developed eggshell thinness threshold-for-effect (for breakage during gestation). Eggs of different bird species vary dramatically in size, shape, and required incubation period (among many other things). Additionally, the weights of nesting parents of various species are likely to be vastly different, and for the hypothetical case, this will almost assuredly be true for the receptor of concern and the surrogate species. Unless we are armed with a precision tool in the form of a rather exacting regression of nesting parent body weights and the required eggshell thicknesses to withstand the weights of nest-guarding parents, we are not equipped to proceed with direct health status assessment.
- The RSA user must always recall that although RSA is rodent-based, terrestrial ERA remedial efforts and concerns to be addressed are not rodent-based and shouldn't be. Sites are simply not remediated (or even considered for remediation) for the purposes of protecting small rodents. The patented RSA method therefore, was assembled to be able to pronounce judgment about the well-being of the larger mammals that contact contaminated sites, and for which we haven't the liberty to study closely.

Before reviewing the second essential point forming the framework of RSA, it is imperative that the reader fully appreciate why the small rodent is such an exceptional choice as the method's monitored organism. The advantages accruing to the choice cannot be overstated. While small rodents may, by default, be the only ones that can be destructively sampled (thereby effectively removing all other terrestrial species from consideration), for all intents and purposes, they are maximally exposed terrestrial receptors. They live in continuous contact with the soil, sporting miniscule home ranges (of the order of 1 acre) that effectively lock them to a given site. They are additionally and necessarily year-round (i.e., non-migratory) species. There should be no objection to using small rodents as sentinels in the mammalian health assessments of RSA applications when they already serve as sentinels for other mammals as part and parcel of the conventional desktop ERA process. It is also true that laboratory studies with mice or rats form the basis of nearly all mammalian SSLs and TRVs used in conventional ERAs (US EPA 2003). RSA theory then, reasons that if small rodents can be used to support the conventional process that does not (even) venture to the out-of-doors to even observe small rodent activity, they can certainly be used to support field-based mammalian receptor health assessments.

The second critical piece contributing to RSA's formation was the recognition that reproduction is the toxicological effect of greatest concern to us. This is evident in a review of toxicity test endpoints spanning multiple classes and genera, the toxicological bases of the TRVs most often employed, and the nature of topical discussions amongst site stakeholders. Importantly, what propelled the assembly of RSA was not simply having arrived at the understanding that a truly utilitarian field-truthing scheme would need to be reproduction-based; along with selecting reproduction as the most worthy endpoint to track, came the recognition also that a reproductive health assessment could necessarily make for an assessment of overall animal health. Thus where reproduction could be demonstrated to be adequate or non-compromised, it could be rightfully concluded by extension, that health in a comprehensive context was also satisfactory. Through so much of the earlier chapters it has been argued that the full-fledged goal of ecological assessment work for contaminated sites is to be able to arrive at a conclusive understanding, that is, one where no additional assessment efforts need be planned. With reproduction appropriately assessed in the field (as with RSA), otherwise anticipated comments and challenges along the lines of "Well, reproduction seems to be all right, but what about ____", could be quelled.

Defining overall health through reproductive health is a central component of RSA theory, and is probably a more all-encompassing concept than many may realize. The concept is first saying that an inattention to other toxicological endpoints can be deliberate because reproduction subsumes the others. (It is worth noting that where traditional ERAs almost solely rely on reproduction, this is not grounded in elements of RSA theory. That ERA has opted to crank out reproduction-based HQs to the near exclusion of HQs that speak to other endpoints, stems from profiling reproduction as being the most sensitive endpoint. True or not true, the great attention reproduction has garnered has served to let ERA practitioners and ecotoxicologists off the hook; these professionals need not set themselves down to the task of developing TRVs and other toxicological benchmarks for additional endpoints. We find the EPA, for example, frequently remarking to the effect that behavioral and neurological effects in ecological receptors are very important to them. With the so-stated high interest, we might ask why TRVs for these endpoints haven't been developed.) That properly structured reproductive assessments subsume other endpoints has a purposeful meaning in RSA. The concept conveys that even if there are biological features or measures in site receptors that widely deviate from the norm (i.e., from what is observed in receptors of the same species at reference locations), these are inconsequential if reproduction is known to be proceeding normally. The decades-old nature of the contaminated sites that submit to ecological assessment comes to bear here. Should reproduction be deemed to be proceeding adequately after so long, it matters not that tissues and organ systems may be affected in any measureable way. If an animal in the field can be shown to be doing what's expected of it, namely to perpetuate its own by producing young, it shouldn't matter that its eyesight or hearing is less than optimal, that its spleen is enlarged, or that its protein metabolism is slowed.

In its early development, RSA saw that a reproductive assessment scheme was all that was needed in order to declare an animal healthy or unhealthy. For mammals at contaminated terrestrial sites, RSA also saw that the essential question to be answered in each case would always be the same – is reproduction compromised in site-exposed animals? Combining much of the foregoing text then, specifically needed was a reproductive testing scheme that involved field rodents. If maximally exposed site rodents were observed to be reproductively sound, it could be concluded through extrapolation (i.e., through noting that all other mammals have far lesser degrees of site contact than do rodents) that the larger mammals of interest that we are unable to get our hands on are also reproductively sound. If some (any) degree of reproductive compromise *was* observed in site rodents, it would be conservatively concluded that the larger mammals of interest were compromised as well. The springboard for the full development of RSA was the stumbling onto a singular and invaluable piece of reproductive biology information, namely that the conventional sperm parameters of count, motility, and morphology – the very same parameters routinely monitored in fertility clinics – are all barometers of reproductive capability in rodents. Previous to the project team's discovery of this fact, it had only been reasonably suggested that just as lowered sperm counts in men can demonstrably contribute to reduced fertility, lowered counts in rodents might similarly act. Those on the earlier-mentioned project team were rampantly naive regarding many facets of sperm parameter science and assessment. They were completely unaware that rodent sperm is assessed daily in hundreds of pure and applied research interests in laboratories across the US and beyond. They had no knowledge that the very computer-assisted sperm analysis equipment used in human reproduction studies is also used in commonplace rodent work, such as occurs in the pharmaceutical industry (albeit with optical software adjustments made to correct for the morphological differences of human and animal sperm). By far, the most meaningful sperm parameter information the project team was to soon acquire was that the degree to which a given parameter would need to be offset in order to impact reproduction, was known. Harnessing this information would allow the user to size up the reproductive capability of mammals actively utilizing sites in the present, precisely what the alternative approach to ecological assessment strives for. With RSA's fundamental underlying premise (identical to that of the more generic "direct health status assessment") understanding critical health effects at aged sites to have already been elicited, the project team would be looking for somatic evidence of reproductive effects, and relinquishing any interest on the possibility of reproductive effects first springing up at some point in the future. Importantly, the field-truthing effort would involve the most sensitive and favored toxicological endpoint and the most highly exposed mammal.

The acquisition of *bona fide* thresholds-for-effect – and for the highly prized endpoint of reproduction, no less – heralded a quantum leap forward in the field of ERA, or more correctly ecological assessment. No longer would comparisons of populations in the lab or field be limited to noting nominal differences between exposed and non-exposed groups (e.g., that liver weights of one were higher than those of the other) and then assuming, for possibly very inappropriate reasons, that certain observed differences are problematic (i.e., health-impacting). A review of the sperm parameter thresholds-for-effect is appropriate at this juncture and will undoubtedly enlighten the reader.

- *To what extent does the sperm count have to be suppressed for reproductive success to be compromised?* An awareness that rodents as a group are robustly fertile, producing 10–20 times more sperm than they need to, facilitates the answer to the question. With excellent assurances, we find that the sperm count needs to be lowered some 80–90 % before we need to think in terms of fewer litters being born (fewer successful matings occurring, as we might think of it) or smaller sized litters being produced (Chapin et al. 1997; Bucci and Meistrich, 1987; Gray et al. 1992; Meistrich et al. 1994). Alertly, this piece of knowledge should be recognized for its potential to incite naysayers to challenge RSA practice on two tracks. If it is firmly established that massive count reductions are needed to compromise reproduction[4], and we encounter a 50% reduction in a field application, naysayers can be expected to forget the established threshold in a heartbeat. Admittedly it is an understandable reaction[5], for our scientific bent causes us to view changes of considerable magnitude a certain way, and particularly where a prime biological function is at stake. Indeed in RSA's early life, the sperm count threshold-for-effect that it utilizes has been scrutinized, although numerous peer-reviewed publications bear out reproductive resiliency

[4] Indeed, there is a documented case of rodent sperm count needing to be reduced 99% before reproductive.

[5] "You mean to say that sperm count at the site can be 50% lower than at the reference location, and there's nothing to worry about?"

where sperm counts are vastly reduced. Perhaps human sperm count declines (of substantially lesser degree) are more devastating, but human health is not the concern here; ecological receptor health is. Whether it is the sperm count threshold-for-effect specifically, or the threshold for one of the other sperm parameters, it is suggested that it is our non-familiarity with having established biological thresholds-for-effect that can make RSA's thresholds hard to accept. Reputedly it is only for the sperm parameters that thresholds-for-effect exist. This means that for the every conceivable piece of biological information that can be collected, man does not know the extent to which the information type must be short-changed before a biological function is compromised. Thus, nerve conduction velocity can be slowed in a bird, but how severely slowed it need be to interfere with wing flapping that allows for lift or sustained flight is unknown. Liver enzyme levels can be reduced, but unknown is the degree to which suppression must occur such that an animal cannot adequately emulsify fats or metabolize its diet in a more overall sense.

Having only discussed the first of the three sperm parameter thresholds, we can think of a second anticipated RSA challenge. We could expect it to be articulated that the RSA method is insufficiently sensitive, in that it requires a dramatic effect to trigger a determination of impact. In essence, the sentiment of the doubtful individual is one of: "But we might not ever find field rodents at contaminated sites to have their sperm counts reduced by some 90%!" Such a sentiment unfortunately reveals being uncomfortable with hearing that rodents (themselves) and the larger mammals for which the field rodent serves as a surrogate, are not health-compromised at contaminated sites. There are many reasons why health effects might be absent at contaminated sites, just one of them being the vast quantities of elapsed time that may have favored receptors outbreeding chemical stressor influences. RSA was invented only to ascertain how site-exposed receptors are faring with regard to the most choice and critical of endpoints. If RSA applications should often – or always – reveal that reproduction isn't compromised, we should come to accept the findings, and not seek opportunities to denounce a clever and viable assessment scheme that reviews the very animals that occupy contaminated properties. The challenge for the naysayer, at least in a context of rodent sperm counts, is to recognize that nature engineered it such that rodents can tolerate substantial impacts before succumbing to lesser reproductive output. For the broader reading audience and with a lesser emphasis on any one indicator of health, there is more to which we need to be open. For the betterment of science and for the management of the contaminated sites that concern us, we must accede to the possibility that instances of impact might not occur, or may only occur highly infrequently.

- *By how much must the sperm motility be reduced to signify reproductive compromise?* The literature is again replete in this area, and the evidence supports reductions in the percentage of properly moving sperm of 40–50% as being needed for fewer and smaller-size litters to be produced (Chapin et al. 1997). To date, it appears that there has been but one instance of exceedance for this threshold that is curiously expressed as a range. In that case, the motility of the *Peromyscus* sp. site population was 47.8% less than that of the reference location, putting the reduction precisely within the threshold's range. As with the sperm count discussion, it may be that mammalian sperm motility is also unlikely to be impacted by a site's chemical stressors to an extreme. The next highest recorded motility reduction in the annals of RSA trials is 25.8%, a figure rather far removed from the threshold-for-effect. It should be noted that for the varied sites that have submitted to RSA, encompassing many states, widely divergent habitats, and with no less than seven rodent species considered, never has the sperm count threshold-for effect been triggered. The reader would again do well to appreciate that as per nature's design, contamination we encounter at sites may not pose sufficient stress to reduce sperm motility to the level of the threshold. Once again, it is not anyone's fault that the percentage of properly moving sperm needs to be effectively halved to draw our attention.
- *Relative to the condition in chemically free rodents in the wild, how much of an increase in the percentage of misshapen sperm need there be to trigger reproductive impacts?* By way of review, every mammal species produces some sperm that are morphologically imperfect, missing either the head or tail segments, or being noticeably bent. These sperm will not go on to fertilize ova, but more importantly, as the percentage of misshapen sperm rises, fewer successful matings occur. The reader might be surprised to know that in all human populations, 50% of sperm produced are morphologically imperfect! In spite of such a statistic, what apparently accounts for the growing world population is the sheer number of sperm in the ejaculate. For our interests here, rodents are radically different from humans in that the percentage of misshapen sperm (hereafter, the control rate) is in the general range of only 2–3%. For a third time, the research is unequivocal regarding a sperm parameter threshold-for-effect (Chapin et al. 1997); an increase of just 4% over the control rate is coincident with compromised reproduction, a statistic that even the savvy biologist would probably not expect. Here, in stark contrast to the sperm count arrangement, we would not anticipate expressed misgivings from naysayers; when the threshold-for effect can be so easily exceeded, who could argue that we are not being sufficiently protective by conducting a comparative sperm morphology review? Sperm morphology then, is apparently the most sensitively set of the sperm parameter thresholds established for

rodents (and by extension, other mammals), in the sense that it needs the least degree of shift in the non-favorable direction (i.e., shifted higher) to translate into compromised reproduction. The review of this parameter for chapter needs would be incomplete without relating what we might term performance scores from the past RSA applications. Here it should first be noted that RSA applications are to be credited for their having supplied the first and only data on sperm motility and sperm morphology in wild rodents of both pristine and contaminated environments. The greatest relative shifts in morphology in RSA history have been ± 0.6% with no cases being statistically significant. As with the discussions on sperm count and sperm motility, this review would also be incomplete without reminding readers that rodents and other mammals might be incapable of achieving chemically triggered relative increases in abnormally shaped sperm of more than 4% above the control rate. It would be worthwhile to note a distinction of import here, however. With regard to count and morphology, the magnitude of the shift needed to confer a finding of reproductive impact is considerable (90% and 40%, respectively) and it seems unlikely that such shifts will occur in the field or be identified. We cannot ignore that by the time we sample rodents (as the most exposed receptors) at contaminated sites, decades of time and hundreds of generations have both passed, with these providing unlimited opportunities to work around reproductive stressors that may present themselves. Aside from these opportunities, in the specific case of sperm morphology, there is clear documentation of this parameter being strongly buffered, i.e., to be incapable, for all intents and purposes, of having its (comparative) threshold exceeded (Tannenbaum et al. 2007). Repeated RSA applications having found site and reference location rodents to be indistinguishable with regard to this parameter, comes to usefully support what has to date been a laboratory rodent-based observation only. It remains a matter for the evolutionary biologist to establish why sperm morphology is so tightly buffered, and RSA utility is certainly not dependent on resolution of this question. In the author's humble opinion and as RSA theory captures it, with sperm morphology needing such minimal skewing to downgrade the essential service of reproduction, nature has opted for the parameter being rather immovable.

The above treatment on sperm parameter thresholds creates the opportunity to revisit standard fare in ERA decision making, something central to the present discussion. Although it wasn't the intention, through the assembly of RSA's technically supported elements, we find that what has effectively come to be doctrine for deciding when a species is health-compromised to have been completely toppled. Invariably for a given biological measure, ERA project teams negotiate a 20% difference in an unfavorable direction as being indicative of ecological impact. When observed in the field, this difference forms the basis for motions to invoke remedial action. A decrease in community biomass of 20% relative to background, a 20% decrease in fin length, or a 20% increase in a metal concentration in the femur, would serve as examples. Vis-à-vis RSA's comparative-context sperm parameter thresholds-for-effect, we would expect biologists to consistently adopt an errant interpretation unless otherwise instructed. Thus a relative 50% reduction in sperm count, considerably exceeding the accepted 20% "standard," would perfunctorily be interpreted as a reproductive system impact of proportion, yet it is no such thing. At the other end of the spectrum, a relative increase in misshapen sperm of only 4% would casually be dismissed since it is so far removed from the 20% figure. Such a minimal increase is, however, fully indicative of high-consequence reproductive impacts. This review exercise comes to consciousness raise; to remind us that there is probably never a time when ERA practitioners are well in touch with the responsiveness of the biological functions that interest them. The exercise is also a clarion call for those who perfunctorily apply the "20% rule", to endeavor to learn of actual thresholds-for-effect. Given (a) the paucity of instances of ecological impact that we are able to cite, (b) the fact that our sites are so often insufficiently large to house enough important receptors, and (c) that sites are decades old before we investigate them, it would be best to suspend application of this "rule."

With the role served by RSA's thresholds-for effect now reviewed, the greater RSA method and its supporting theory can be operationally set forth. RSA is, of course, not a risk assessment method; it does not express the likelihood of ecological receptors developing toxicological endpoints. No method can do such a thing, but more than this, the interest at contaminated sites should be in seeing if effects of interest have come about, and particularly because they have had so much time to do so. RSA applications, in addition to not being risk assessments, are not toxicological, experimental, or research endeavors involving animals. RSA applications therefore, do not involve (a) the purchase of laboratory-reared animals from suppliers; (b) randomizing animals into treatment groups; (c) assuming animal housing tasks such as adjusting diets, cleaning cages, and controlling indoor lighting and temperature; (d) purchasing chemicals from suppliers; or (e) administering chemicals as part of a dosing regimen. In stark contrast to toxicological investigations that support conventional ERA, RSA applications harvest animals from their natural settings, record diagnostic information from the animals that relates to ERA's toxicological endpoint of greatest concern (i.e., reproduction), and compare the recorded measurements to scientifically established and technically defensible standards. RSA's evaluative process can be compared to the case of a doctor reviewing the blood work for a new patient who had recently scheduled a well visit. For a patient who presented with a fasting blood glucose level of 350 milligrams per deciliter (mg/dl), the doctor will necessarily conclude that the patient has a compromised blood sugar metabolism, and that the patient is diabetic. The determination in that case follows from the availability of a well-established "norm" for blood glucose levels in fasted individuals and the knowledge that considerable exceedances of the "norm" describe a diabetic condition. The RSA method is similarly able to render its useful reproductive capability determinations for mammals at contaminated sites because of the availability of the established sperm parameter-based thresholds-for-effect that are used in a comparative assessment scheme.

RSA is necessarily a top-down approach. It intends to evaluate site receptors as they appear to us, allowing us to decide on whether they present with illness that presumably stems from site chemical exposures. The effort to craft RSA as a top-down approach was not only spearheaded because we have the capability to collect representative site receptors and directly evaluate them. The effort proceeded in recognition of bottom-up approaches being unworkable. Modeling the assumed chemical exposures of site receptors through food chain transfers carries with it a host of inaccuracies (Thiessen et al. 1997), in addition to yielding information of a type that doesn't facilitate decision making. The open-minded reader should recall (from Chapter 2) that the be-all and end-all of ecological assessment is not in estimating how much more than a safe-level or effect-level dose an animal is thought to be ingesting. Further, the least of the problems with such an approach is that it is likely to leave in its trail the question of why we see receptors at sites at all (i.e., since the receptors are, so it seems, consuming unhealthful chemical quantities, why haven't they yet succumbed?).

RSA as a top-down approach differs most from bottom-up assessment efforts in its holistic considerations. Whereas bottom-up modeling will forever report its meager findings in a chemical-by-chemical manner and with only the ingestion pathway considered, RSA has no need to make such major and far-reaching concessions. With RSA, the receptor of concern is recognized for its functional role as the great integrator of contaminant inputs, both in terms of processing multiple chemicals at a time, and with multiple (i.e., three) uptake routes operating concurrently. In looking to see if a critical biological function has been compromised and where prominently, the exhaustive universe of chemical inputs has been considered, we discover that the chemical mixtures riddle (Borgert 2004) has been resolved for all intents and purposes. We discover too, that with direct health status assessment (as in RSA), uptake routes no longer have to be assessed discretely. Our inclination should always be to assess the combined effects of all uptake routes, for to our knowledge, the actual site receptor hasn't the capability to voluntarily take a given uptake route offline. The holistic nature of RSA assessments triggers several finer points of discussion. First, if the objective is to pass judgment on the health status of an appropriately selected site receptor, we must be certain that the biological functions under review are appropriate ones. We return to considerations described in Chapter 4. For the same level of effort that a standard RSA application entails (thinking primarily in terms of live-trapping small rodents), one might opt to comparatively assess cardiac efficiency at contaminated sites and their habitat-matched reference locations. Presumably a site's chemical suite could act to reduce cardiac efficiency, and presumably as well, we could assemble the equipment and the procedures for recording the sensitive readings that inform in this area. Unless we are privileged to know the extent to which cardiac efficiency metrics need to be altered in order to signify that health overall or that life expectancy is compromised, the measurements hold no utility. To the not fully RSA-affiliated, RSA could be carelessly captured as "that method where sperm counts are compared in rodents from two nearby locations." Not explaining why the sperm count and the other sperm parameters are recorded, could well give rise to those desirous of establishing their own personal mark on the ERA landscape, electing to gather some esoteric somatic measurement (perhaps left uterine horn diameter proximal to the urinary bladder) to be compared. To underscore the point, it is only for the standard three sperm parameters that we can speak in terms of biologically significant differences.

What chemicals can cause sperm parameter shifts? This is a question that actually bears no practical relevance to RSA (in terms of such things as deciding whether or not to apply the method), but speaks tomes in a context of RSA theory, a distinction to be clarified in the next paragraphs. To be sure, there are chemicals that are known spermatoxins and these include quite a number of metals and an assortment of organic compounds. Independent of the specific chemicals that occur in the environmental media with which terrestrial receptors interact though, RSA theory recognizes that our greatest concern (or perhaps, fear) is that site receptors today might be reproductively comprised.[6] Receptors then, at each and every site and independent of a site's contaminant suite, need to have their reproductive capability assessed no matter what. Well-meaning ERA practitioners could wander far astray by electing to only apply RSA after certifying the presence of at least one *bona fide* spermatoxin within a site's contaminant list. Such a myopic approach would overlook several points. First, not every naturally occurring or synthetic chemical has been sufficiently researched for its ability to impair sperm. Additionally, spermatoxicity might only arise from the combined actions of chemicals, some or none of which might presently be understood to be spermatoxins. Potentially then, there is much that we could be missing out on by not assessing reproduction (in a novel and direct way, no less) at any *bona fide* site. Hence in RSA theory, it is enough to simply have chemicals present to trigger a reproductive capability analysis. As was loosely discussed in Chapter 4, when ERA practitioners elect to run toxicity tests of any sort, there is

[6] ... and not that they *might become* reproductively compromised.

no preliminary screening step whereby a medium's contaminant suite is reviewed to ascertain the presence of at least one chemical that has been shown to interfere with a test's endpoint. In the annals of ERA investigation, contaminant suites in relevant media were never first reviewed before it was decided to run toxicity tests involving plants, earthworms, amphipod or copepod crustaceans, or fish. We might think about this concept in a different way: assuming testing isn't prohibitively expensive or unwieldy to conduct for some reason, we should want to know if a site is impinging on plant or animal health, and particularly reproduction. Further, if we *didn't* run the tests that are available to us (where these are defensibly utilitarian, of course), we would be remiss and we would be deserving, too, of certain stakeholder clamoring because of our oversight.

In addition to the good sense it makes to apply RSA at every contaminated site because of the method's ability to so well inform, we should note that contaminant screening efforts would invariably lead us to run the method anyway. Naturally occurring inorganic species such as aluminum, arsenic, cadmium, chromium, lead, mercury, and nickel are established spermatoxins that impinge on one or more of the attention-drawing sperm parameters (Al-Omar et al. 2000; Chowdhury et al. 1989; Cody et al. 1981; Dilley et al. 1982; Evenson et al. 1989; Homma-Takeda et al. 2001; Levine et al. 1984; Li et al. 2001; Linder et al. 1992; Pandey et al. 1999; Pandey & Singh 2002; Pant et al. 2001; Reddy et al. 1996; Xu et al. 2001). At this juncture the reader should understand that physical stressors are also commonly at play at contaminated sites. Although perhaps not yet proven, it is reasonable to assume for example, that noise and land vibration attributable to repetitive high energy munitions detonations, can impose stress to the point where reproduction is compromised, even taking the form of altering sperm. More often than not it may be that a significant non-chemical stressor (e.g., an unusual or extreme weather pattern) will be present, and by undertaking a specialized reproductive review like RSA, we run the risk of wrongly ascribing a compromised reproductive condition to site chemistry. We should nevertheless take this risk, for again the interest in ecological assessment is (or should be) in knowing if a site has harmed receptors. Colloquially we may say that the top priority with RSA is in finding out if reproduction has been short-changed, and that we can leave the worry about effect causation for later. In fact this is distinctly what RSA is about, and it leads us to a consideration of RSA outcomes. As with the general model of direct health status assessment reviewed in the previous chapter, RSA was not designed to pinpoint the stressor or stressors responsible for one or more sperm parameter thresholds being exceeded. Further, the very nature of top-down assessments for field-exposed animals rather handily precludes such causative factor identification. In that case where one or more sperm parameter thresholds were found to be exceeded, chances are that skilled environmental biologists could soon enough forensically determine the offending agent or agents.

Can RSA sperm parameter thresholds be exceeded? The question quickly morphs to: Can any vital biological function be compromised in site-exposed receptors? Critics and naysayers will resent the latter question, but for ecological science to proceed, it must be faced head-on. RSA theory would answer "no" to both questions, basing this on the RSA method's fundamental underlying premise. The theory is first open to the possibility that health impairments of one sort or another do arise, presumably in the years shortly after a site becomes contaminated. A failure to detect health impairment for the investigative mode under which we operate though (i.e., assessments conducted decades after contaminant release events occur), could well mean that an effect that did develop, has since disappeared owing to any of a number of corrective mechanisms. Alternatively, a failure to detect health impairment can mean that even in the early years post chemical release, site receptors never succumbed to a condition at all. The basis for this is straightforward; sites don't become radically contaminated overnight. To the contrary, chemical release, whether by deliberate dumping or application, seepage, regular deposition to the air, etc., occurs very gradually. Consequently ecological receptors in the wild have the opportunity to adjust to chemical stressors that slowly phase in, perhaps never experiencing a sufficiency of chemical stress to have a toxicological response elicited.

The attentive reader should note that more than just theory has been applied to the question of sperm parameter threshold inviolability. The developers of RSA have every interest trying to "force a failure"; that is, to seek out sites that are in one way or another exceedingly contaminated. What though, if even these select sites fail to produce a sperm parameter threshold exceedance? Must we still reserve judgment on RSA method capability? Again, it would be unfair to suggest that because computed HQs for the exceedingly contaminated properties are greater than 1, while RSA applications at the properties do not record sperm parameter exceedances, RSA is failing to detect problems that are actually present. The anticipated HQs for the so-described extremely "hot" sites would undoubtedly be off-scale. Where some would be quick to conclude that the RSA method puts in its worst performance for sites with excessive contamination, the reader should sooner recall that excessive HQs are meaningless impossibilities. Whether NOAEL- or LOAEL-based, no organism on our planet can withstand 3 or 4-digit HQs or still higher ones, and even NOAEL-based HQs of 20 are suspect (Tannenbaum et al. 2003b). A HQ review cannot resolve the current matter of why there has yet to be a case of RSA sperm parameter threshold exceedance. As has been suggested before and in line too with RSA's fundamental underlying premise, animals in the wild with decades of site exposure to their credit, might simply be more resilient to the toxic effects of chemicals than we can imagine. There are also the matters of (a) our not wanting to believe that ecological receptors can go about their business living amid contamination, and (b) our unwillingness to hear that the HQ approach to ecological assessment has been bested.

RSA developers have one other test situation in mind to resolve the threshold inviolability matter, namely to conduct RSA at sites that have only (relatively) recently become contaminated. Although we never find sites that are perhaps five years old or less submitting to ERAs of any form, conceivably there hasn't been sufficient opportunity at these for site receptors to adjust to the contamination. Who knows what the future has in store with regard to additional forthcoming information that might further support the abilities of RSA?

RSA was not invented because anyone knew its test system to be most unlikely to reveal that impacts were apparent. It was invented with an open mind, with the willingness to see for the thresholds-for-effect we are so fortunate to have, if the most sensitive toxicological endpoint can be reached or exceeded. For the user then, there are alternative ways to understand or interpret RSA's outcome universe, one which is completely free of cases of threshold exceedance. On the one hand, the collective data can be seen to be clearly conveying that, terrestrial ecological receptors (at a minimum, mammals) are beyond toxicological reproach. On the other hand, the test outcomes stand to be scrutinized beginning with the well-anticipated challenge of: "What good is RSA if it can't ever identify an instance of reproduction having been compromised?" Importantly, the inability to identify a case of reproductive failure needn't reflect RSA method weakness. Reasonably the absence of cases to cite reflect the contaminant suites and contaminant concentrations we have encountered since the Superfund era arrived, not being as threatening as some might expect. It's no one's fault if the contamination isn't as bad as was thought, or that ecological receptors in the wild are more resilient than we would believe. Sooner than mount attacks on a method that has the potential to fully re-orient and re-energize the field of ecological assessment, ERA practitioners should take to heart the implications of the novel RSA work that has been conducted. And so, is the ERA practitioner fully open to the possibility that we may never come across cases of ecological receptor harm – be it in rodents utilized in RSA applications or in any other context? Those who are not so open, are duly challenged to cite any instance at all of ecological impact, recalling that HQs >1 are far from indications of impact.

Is RSA the answer to ERA? Speaking for the terrestrial investigations that consume us, and for mammals (at this time), one of only two terrestrial groups upon which we focus, reasonably it seems so. The method is certainly more efficacious than HQ computation which is but a screening tool that principally models soil chemical concentrations, is capable only of assessing chemicals singly, and in all truthfulness commences with a flawed question ("What might happen to site receptors if they continue to be exposed?"). If we can lay down our bias and our unwillingness to consider an alternative assessment approach, we can discover that the needed scientifically defensible supports are in place for RSA. For a minimal investment of time in the field, and for no more cost than an ERA of boilerplate design – a textual treatment only – we can let the very specimens that we procure from contaminated sites inform us, and inform us well. There is life beyond the HQ if we are willing to recognize it.

RSA is the answer to ERA because it so vastly improves our assessment capabilities. It effectively resolves the chemical mixtures riddle, it accounts for all three routes of chemical uptake, and it accounts for physical site stressors that heretofore have not been addressed in conventional ERAs. RSA can be the answer to ERA only if we are willing enough to break with traditional procedures, if we can be willing to hear that animals in the wild might not be experiencing any biological health deficits, and if we can cease insisting that all contaminated sites necessarily submit to remediation.

Most in order at this point is a brief yet comprehensive and synthesized concluding review of the alternative approach to understanding ERA that has been set forth earlier. It is intended to be instructional, and it is presented with the hope that it be given due consideration at all ranks.

There is no way to assess ecological risk on planet earth – a basic fact and not an opinion. It is also a most difficult pill to swallow; those who have assembled the process that we continue to employ and those who routinely apply it, surely don't want to hear that the process doesn't work and that the process isn't needed. It doesn't matter though, that we haven't a risk assessment process for ecological receptors situated at hazardous waste sites and the like. By the time we begin to investigate contaminated terrestrial and aquatic sites, far too much time has passed to make it reasonable to speak in terms of defining the *potential* for receptors to develop incapacitating effects brought on by their exposures. This is not to say that we should shun all forms of assessment. In place of predicting or forecasting health effects that could come due, our energies should be directed, with certain qualification, at conducting assessments that can identify health effects that have already arisen in the actual receptors dwelling at the contaminated sites. The only somatic or behavioral changes worthy of investigation should be those that bear on survival and the capability to play out essential biological functions. Further, only those measurable changes for which biological thresholds-for-effect exist, should be studied. Disinclined as ERA practitioners may be to leave the comforts of the office and desktop to gather receptors in the field within prescribed biological windows of opportunity, such is the way to proceed; we should not be interested in the biological responses of laboratory-reared specimens, and we should not be comparing such responses to those of wild-type specimens. Even before embarking on direct health status assessments in lieu of ineffective HQ and EEQ computation, spatial scale considerations must come into play. While scale is presently all-the-talk (Kapustka 2008), consistently overlooked is the fact that sites do not support a sufficiency of receptors (that truly concern us) to make any form of assessment worthwhile. This arrangement is simply a function of the natural world order; contaminated sites are relatively small, and the species that concern us are sparsely distributed in nature and have relatively large home ranges. This is a wonderful mix. There is no shame or embarrassment in reporting that we've no need for assessment either because there aren't enough receptors exposed, or because receptors insufficiently contact sites such that they would come to present with illness. There is no obligation to compute quotients at each and every site as a matter of course. In those instances where we compute them nevertheless and find them to be acceptable, we delude ourselves into thinking that they are informing us that site receptors are protected. We could have and should have instead, had our ecological assessments take the form of one-paragraph summaries indicating that formal analysis wasn't called for altogether.

Direct health status assessment opens up a world of opportunity – to a) discover if *actual* site receptors are *actually* health-compromised, and b) to fully secure the understanding that simply because contamination abounds and lingers in the environment, this need not translate into health effects setting in (even in maximally exposed forms). Direct health status assessment has already taken hold in the form of the patented Rodent Sperm Analysis (RSA) method (US APHC 2009). For the few terrestrial sites that are legitimately large (i.e., to a point where a sufficiency of larger species can be expected), it can provide as definitive proof as is possible that mammals are/are not sufficiently protected. Although RSA has received its share of accolades from the US EPA, for several reasons it is unlikely that this agency will formally adopt the method or champion its use. First, the US EPA didn't devise it, and a high-ranking agency doesn't like seeing its work bested. Additionally, reigning regulatory authorities don't take well to reform, and the RSA method, although consistent with ERA doctrine, ventures into a new and forward-thinking dimension. Finally, RSA has the great potential to show that terrestrial sites are commonly protective of the biological forms they cater to, and among ERA practitioners, there is great disinclination to hear such good news. Seemingly by virtue of having sought out this book, the reader is open-minded and genuinely interested in seeing improved science infiltrate the field of ecological assessment. It is hoped that the reader will resist falling prey to the biases that presently pervade our field, and be empowered to pursue alternative assessment schemes where it is appropriate to do so.

References

Al-Omar, M.A., Abbas, A.K., & Al-Obaidy, S.A. (2000) Combined effect of exposure to lead and chlordane on the testicular tissues of swiss mice. *Toxicology Letters* 115:1-8.

Borgert, C.J. (2004) Chemical mixtures: An unsolvable riddle. *Human and Ecological Risk Assessment* 10:619.

Bucci L.R. & Meistrich, M.L. (1987) Effects of busulfan on murine spermatogenesis: cytotoxicity, sterility, sperm abnormalities, and dominant lethal mutations. *Mutation Research* 176:259-268.

Chapin, R.E., Sloane, R.A., & Haseman, J.K. (1997) The relationships among reproductive endpoints in Swiss mice, using the Reproductive Assessment by Continuous Breeding database. *Fundamentals of Applied Toxicology* 38:129-142.

Chowdhury, A.R., Makhija S., Vachhrajani, K.D., & Gautam, A.K. (1989) Methylmercury- and mercuric-chloride-induced alterations in rat epididymal sperm. *Toxicology Letters* 47:125-134.

Cody, T.E., Witherup, S., Hastings, L., Stemmer, K., & Christian, R.T. (1981) 1,3-Dinitrobenzene: Toxic effects in vivo and in vitro. *Journal of Toxicology and Environmental Health* 7:829-848.

Dilley, J.V., Tyson, C.A., Spanggord, R.J., Sasmore, D.P., Newell, G.W., & Dacre, J.C. (1982) Short-term oral toxicity of a 2,4,6-trinitrotoluene and hexahydro-1,3,5-trinitro-1,3,5-triazine mixture in mice, rats, and dogs. *Journal of Toxicology and Environmental Health* 9:587-610.

Evenson, D.P., Janca, F.C., Baer, R.K., Jost, L.K., & Karabinus, D.S. (1989) Effect of 1,3-dinitrobenzene on prepubertal, pubertal, and adult mouse spermatogenesis. *Journal of Toxicology and Environmental Health* 28:67-80.

Gray, L.E., Marshall, J.O., & Setzer, R. (1992) Correlation of ejaculated sperm numbers with fertility in the rat. *Toxicologist* 12:433

Homma-Takeda, S., Kugenuma, Y., Iwamuro, T., Kumagai, Y., & Shimojo, N. (2001) Impairment of spermatogenesis in rats by methylmercury: involvement of stage- and cell- specific germ cell apoptosis. *Toxicology* 169:25-35.

Kapustka, L. (2008) Limitations of the current practices used to perform ecological risk assessment. *Integrated Environmental Assessment and Management* 4:290-298.

Levine, B.S., Furedi, E.M., Gordon, D.E., Lish, P.M., & Barkley, J.J. (1984) Subchronic toxicity of trinitrotoluene in Fischer-344 rats. *Toxicology* 32:253-265.

Li, H., Chen, Q., Li, S. et al. (2001) Effect of Cr(VI) exposure on sperm quality: human and animal studies. *Annals of Occupational Hygiene* 45:505-511.

Linder, R.E., Strader, L.F., Slott, V.L., & Suarez, J.D. (1992) Endpoints of spermatotoxicity in the rat after short duration exposures to fourteen reproductive toxicants. *Reproductive Toxicology* 6:491-505.

Meistrich, M.L., Kasai, K., Olds-Clarke, P., MacGregor, G.R., Berkowitz, A.D., & Tung, T.S.K. (1994) Deficiency in fertilization by morphologically abnormal sperm produced by *azh* mutant mice. *Molecular Reproduction and Development* 37: 69-77.

Pandey, R., Kumar, R., Singh, S.P., Saxena, D.K., Srivastava, S.P. (1999) Male reproductive effect of nickel sulphate in mice. *Biometals* 12:339–346.

Pandey, R. & Singh, S.P. (2002) Effects of molybdenum on fertility of male rats. *Biometals* 15:65–72.

Pant, N., Kumar, R., Murthy, R.C., & Srivastava, S.P. (2001) Male reproductive effect of arsenic in mice. *Biometals* 14:113–117.

Reddy, T.V., Daniel, F.B., Olson, G.R., Wiechman, B.H., & Reddy, G. (1996) Chronic toxicity studies on 1,3,5-trinitrotoluene in Fischer-344 rats. US EPA, Cincinnati, OH. ADA315216. US Army Medical Research and Materiel Command, Frederick, MD.

SAIC (1999) Draft Final Phase II Remedial Investigation Report for the Winklepeck Burning Grounds at the Ravenna Army Ammunition Plant, Ravenna, Ohio. Science Applications International Corporation.

Tannenbaum, L.V., Bazar, M., Hawkins, M.S. et al. (2003a) Rodent sperm analysis in field-based ecological risk assessment: pilot study at Ravenna army ammunition plant, Ravenna, Ohio. *Environmental Pollution* 123:21–29.

Tannenbaum, L.V., Johnson, M.S., & Bazar, M. (2003b) Application of the hazard quotient in remedial decisions: A comparison of human and ecological risk assessments. *Human and Ecological Risk Assessment* 9:387–401.

Tannenbaum, L.V. Thran, B., & Williams, K.J. (2007) Demonstrating ecological receptor health at contaminated sites with wild rodent sperm parameters. *Archives of Environmental Contamination and Toxicology* 53:459–465.

Thiessen, K.M., Hoffman, F.O., Rantavaara, A., & Hossain, S. (1997) Environmental models undergo international test: The science and art of exposure assessment modeling were tested using real-world data from the Chernobyl accident. *Environmental Science and Technology* 31:358A–363A.

US APHC (2009) Rodent Sperm Analysis. US Patent 7,627,434. US Army Public Health Command.

US EPA (2003) Guidance for Developing Ecological Soil Screening Levels (Eco-SSLs). OSWER Directive 92857–55, Washington, DC. US environmental Protection Agency.

Xu, L.C., Wang, S.Y., Yang, X.F., & Wang, X.R. (2001) Effects of cadmium on rat sperm motility evaluated with computer assisted sperm analysis. *Biomedical and Environmental Sciences* 14:312–317.

Index

Alternative Ecological Risk Assessment: An Innovative Approach to Understanding Ecological Assessments for Contaminated Sites, First Edition. Lawrence V. Tannenbaum.